The 'mountain' is the highly magnified surface of a crystal of aluminium; magnification × 35 000. (From *The World through the Electron Microscope*, by courtesy of JEOL Ltd.)

Surfaces

H. B. GRIFFITHS

Professor of Mathematics, University of Southampton

Second edition

CAMBRIDGE UNIVERSITY PRESS

CAMBRIDGE

LONDON NEW YORK · NEW ROCHELLE

MELBOURNE · SYDNEY

Published by the Press Syndicate of the University of Cambridge
The Pitt Building, Trumpington Street, Cambridge CB2 1RP
32 East 57th Street, New York, NY 10022, USA
296 Beaconsfield Parade, Middle Park, Melbourne 3206, Australia

First published 1976
Second edition 1981

Printed in Great Britain by
Spottiswoode Ballantyne Ltd,
Colchester and London

British Library cataloguing in publication data
Griffiths, Hubert Brian
Surfaces – 2nd ed.
1. Topology
2. Surfaces
I. Title
514 QA611 80-49884

Second edition
ISBN 0 521 23570 7 hard covers
ISBN 0 521 299772 paperback
(First edition ISBN 0 521 20696 0)

To the memory of Joe Griffiths (1964–70), who loved to make models of all kinds, but especially paper Origami ones.

All royalties from this book are the property of the Cystic Fibrosis Research Foundation, 5 Blyth Road, Bromley, Kent BR1 3RS.

Contents

Preface *page* ix

1 How to make surfaces, and talk about them 1
1.1 What is a surface? 1
1.2 Two rules for making paper surfaces 2
1.3 Some names for things 7
1.4 Some simple surfaces with boundary 9
1.5 Thinking about a Moebius band 10
1.6 Where have we got to? 12

2 Making complicated surfaces 14
2.1 Planar regions 14
2.2 Adding an ear 14
2.3 Non-planar surfaces 15
2.4 Recognizing a twist 16
2.5 Adding a bridge 17
2.6 The Euler number 18
2.7 The punctured torus 20
2.8 The Klein bottle and the projective plane 21

3 Multifarious complications 24
3.1 Building up and breaking down 24
3.2 Multi-bridges and multi-ears 24
3.3 The punctured sphere with g handles 26
3.4 The closed sphere with g handles 27
3.5 Orienting a surface 28
3.6 Important questions 30

4 Families of surfaces 32
4.1 Laminas and cocoons 32
4.2 The annulus family and its relations 35
4.3 The families of planar regions and of spheres with handles 39
4.4 The general plan $\mathscr{S}_{p,q,r}$ 43
4.5 The Trading Theorem 46
4.6 Adding a lamina to $\mathscr{S}_{p,q,r}$ 48
4.7 Rules for recognizing surface families 54

5 Completion of the census of surface families 57
5.1 Some algebra 57
5.2 The Fundamental Theorem 58
5.3 Calculation of p, q and r 59
5.4 Uniqueness of the plan of a surface 60
5.5 The family of a closed surface 61
5.6 Dependence on the order of assembly 62
6 Combinatorial invariants 63
6.1 Repanelling a paper surface 63
6.2 Paper complexes 64
6.3 Triangulations 66
6.4 The Invariance Theorem for a lamina 68
6.5 Refining a panelling 71
6.6 The Euler numbers of a panelling and a refinement 72
7 Order of assembly, and orientability 75
7.1 Mystery surfaces 75
7.2 The Assembly Theorem 76
7.3 The Orientability Theorem 78
7.4 Proof of the Orientability Theorem 79
7.5 Calculation of the orientability number 80
8 Morse Theory of a paper surface 82
8.1 The mountaineer's equation, and Physical Geography 82
8.2 The mountaineer's equation for a general surface 84
8.3 Proof of the mountaineer's equation 90
9 Miscellaneous exercises 95
Appendix A Mathematical theory of surfaces 99
A.1 Notation 99
A.2 Jordan and Schoenflies 99
A.3 Mathematical surfaces 99
A.4 Families 100
A.5 Mathematical language 101
A.6 Further outlook 101
Appendix B Teaching notes 103
Appendix C Hints and solutions to exercises 109
References 124
Index of symbols 125
Subject index 126

Preface to the first edition

Over several years, I have tried to teach topics from the geometry of surfaces to various audiences of adults. These varied in technical skill, from Extra-Mural students with plenty of interest but only old-fashioned arithmetic, to Honours mathematics students with A-level skills and attitudes. Almost all students, however, seemed to find difficulty in thinking about spatial objects, and there was little that they could read, to help them; for the conventional developments of the theory are usually found in advanced texts, and often not in English.

It is also nowadays quite possible to pass public examinations in mathematics while avoiding any 3-dimensional problems and the consequent need for sketching. We then have the cycle: children uneducated in 3-dimensional thinking, become teachers without skill in 3-dimensional thinking, who then leave their pupils uneducated in such thinking, who then become teachers . . .

This book is an attempt to break that cycle, being designed for students who might later teach children. It has two aspects, one mathematical, and one educational. Mathematically, it expounds the topology of compact surfaces as far as the Classification Theorem, and treats also the Morse Theory of such surfaces. In fact the whole treatment is a 'handle–body' approach. However, a conventional mathematical treatment would begin 'logically' with topological spaces, continuity and homeomorphisms, before settling down to combinatorial ideas. Any student who could not jump the initial conceptual thresholds would then be cut off from the rich intuitive material that underlies the later work. And it is the rich intuitive material that a future teacher will need, for introducing children to 3-dimensional thinking, so that he can stress the *meaning* rather than the syntax. (The importance of meaning, as against syntax, is emphasized by René Thom in his lecture to the 1972 International Congress on Mathematical Education: see the Proceedings (*Developments in Mathematical Education*, edited by A. G. Howson, Cambridge University Press, 1973), especially p. 206.)

Consequently, this book is written within the discipline of Mathematics Education, and along the lines suggested in my Invited Address to the International Congress of Mathematicians, Nice, 1970. Certain 'Pedagogical Axioms' are assumed at the outset (and given here the more pleasant names of 'Agreements') and a theory is developed from them. The development uses a language which departs from conventional English as little as possible, but which is isomorphic to a 'pukka' mathematical language.

Indeed, in the course of the development, we *introduce* the notions of definition, theorem and proof, and especially the method of mathematical induction. All these ideas are basically new to the typical student who has covered a traditional A-level syllabus in mathematics, even though he may have heard the words. It is unfortunate that many mathematics lecturers assume that because the student has heard the words, they are a part of his being; and that therefore he will immediately comprehend a lecture course of the austere 'Definition, Theorem, Proof' kind without ever knowing why mathematicians invented such a style. Such professional mathematicians will perhaps appreciate the point of view of this book (even if they do not sympathize with it), if I say that it is written as a corrective to the young man who, having been asked to provide a course on surfaces, lectured on the combinatorial theory of n-manifolds, and in the last lecture said, 'Now put $n = 2$, and we get, trivially . . .'. (The adolescent word 'trivially' is not used in our treatment.) The point is that the young man was interested only in getting the mathematics straight in his own head, not with communicating with an audience and taking into account its mathematical skills and understanding. It is when the mode of communication relates the mathematics to such constraints that we meet the discipline of *mathematics education*. Of course, had the young man's constraints included one that said all his audience were future Ph.D. candidates in mathematics, his own curriculum might have been suitable.

For any reader who may have passed the conceptual thresholds mentioned above, I have included Appendix A, containing an outline of a full mathematical treatment, in which (of course) the Pedagogical Axioms are modelled and proved within conventional topology.

In fact, our treatment is related to Topology, as Synthetic is related to Analytic Geometry; the undefined terms of Euclidean Geometry correspond to real-life points, lines etc., but the development of Euclidean Geometry can be modelled isomorphically within Analytic (Coordinate) Geometry. One does not need to understand the construction of the real number system from the integers in order to grasp the theorem of the Nine-Point Circle, for example, or Pascal's Theorem in Projective Geometry. In the same way, the 'panels' and 'paper surfaces' of our treatment have real-life counterparts, and their mathematics can be discussed in a language familiar to beginners. The material of Appendix A might well be discussed in a course for Final Honours undergraduates; but the bulk of the text is designed for less advanced students. A second Appendix consists of 'Teaching Notes', in which reasons are given for adopting certain strategies (marked with a symbol Ⓣ in the text) – why, for example, we might avoid saying that something is an equivalence relation (because it might involve a student in a large effort of understanding, without a compensatory gain in his geometrical insight). Naturally, if a student already understands the notion of equivalence relation, this understanding can be reinforced when he* meets another example of it, especially if *he* points it out rather than being told.

My hope is that some teachers (actual or potential) will use some of the material of the book for teaching students of various ages. Their professional judgement will tell them which topic is suitable for which student, and they

* Throughout, 'he' means 'he or she' when we refer either to teacher or to pupil.

will pick the conversational style that is best suited to communicate with him. For example, now that in Britain abolition of the 'eleven-plus' examination allows imaginative mathematics to be regarded as the norm in our primary schools, young children might model specific surfaces and work out their Euler numbers. In secondary schools, children can work at that part of the material requiring algebra of the form '$\chi = n$' rather than '$\chi = -21$'. It seems to be necessary to encourage habits of modelling and drawing fairly early, because undergraduates seem to think it undignified to begin practising at their late age. I hope, too, that these studies can be associated with geometrical drawing, and painting and sculpture; and especially that by visits to art galleries, objects may be touched and sensed in the round, rather than being diagrammed or photographed. In this way, pupils may be led to speak their own thoughts on geometry, and not merely be told the thoughts of others. By rational discussion they may be led to appreciate both the civilizing notion of proof and the practical benefits that earlier geometry has brought us in engineering, architecture and science.

The only technical prerequisites needed of the reader are that he can follow a diagram, and that he is not afraid of simple algebraic expressions. In this connection, he should understand that if we have a set containing n objects, then we can refer to them as $P_1, P_2, \ldots, P_n$ (short for the object labelled '1', object labelled '2', and so on up to the object labelled 'n'), and read 'P sub 1, P sub 2, etc., up to P sub n'. A few references are made in the text to other books that may be consulted; for example Hilbert and Cohn-Vossen [9] refers to the book listed as number 9 in the references at the end.

Of course, the reader needs something else, which he as a teacher may possess, but which his intended students may not. I refer to curiosity and the wish to know, as well as the patience and stamina necessary to find answers in a systematic way. It should not be assumed (although it often is, in institutions of 'higher' education) that pupils possess such qualities: certainly young children have curiosity, but pressures of their environment may well discourage patience, and emphasize quick returns. Consequently it is part of a teacher's job to encourage these qualities in pupils, and I hope that I have at least provided some interesting material to help in this encouragement.

Many exercises are included, in the hope that the reader will try them, to increase his insight. The more technical ones are indicated by a star, and are not essential for an understanding of the text; but the reader should not be scared by their symbolism, and he should come back to them when his desire to know inspires him to penetrate the algebra. Some hints and solutions are to be found in Appendix C, but some exercises are designed as 'investigations' and it would spoil them if one commented too much; this is why not all exercises are supplied with hints *or* solutions.

I wish here to thank several friends who made helpful comments on a preliminary draft of the book, and gave warm encouragement. Especially, too, I wish to thank the members of the Cambridge University Press for their advice, and their artist who drew the diagrams.

H.B.G.

Southampton, May, 1974

Preface to the second edition

Several reviewers and correspondents have shown a warm interest in the book and made suggestions for improving it; where possible I have altered the original text accordingly. The principal technical problem has been to give more explanation for identifying a twisted bridge, enabling certain proofs to be clarified or corrected. Sometimes I have been asked for fuller commentary, or for greater precision, especially when first introducing an idea, but here I have not always acceded to the request: my aim has been to allow beginners to get to the heart of the matter without being obstructed by too many written words or minor exceptions. (In a classroom, explanations and exceptions can be given conversationally, but they might well freeze if set down on a page.) Thus, some critics have not fully appreciated the philosophy of my earlier Preface and the Teaching Notes, and their criticism has been to demand that the exposition should have the finished look of conventional contemporary mathematics with every objection prepared for in advance. My point here has been to *generate* objections, in order to create a wish in students for greater precision in language and rigour *once they have some interesting mathematics to care about.* This repeats a historical process so beautifully captured by Lakatos, in his book *Proofs and Refutations* (Cambridge University Press, 1976), where his principal example is the development of Euler's Theorem ($\chi(\text{sphere}) = 2$) – although his thesis is exemplified in Analysis just as well. The reader interested in a fuller discussion of the philosophy behind the present exposition may care to read the interchange between R. Schwarzenberger and myself in the *Mathematical Gazette* (**60**, 1976).

Particular thanks are due to Benno Artmann and Gwylym Edmunds, for their comments after using the text with students; Gwylym kindly allowed me to use his tutorial sheets and examination papers on which to base the extended set of Exercises forming the newly-added Chapter 9. I am also grateful to Tony Gardiner and John Rigby for sending me detailed letters of criticism, and to the translators of the French and German editions for raising fruitful expository points and finding errors. Once again, members of Cambridge University Press gave me advice and help with their usual high standard of expertise and kindness.

H.B.G.

Southampton, 1980

1. How to make surfaces and talk about them

1.1 WHAT IS A SURFACE?

What do *you* mean by a Ⓣ surface? In real life we are surrounded by surfaces – those of furniture, tools, utensils, buildings, fluids, our bodies – and yet most people have surprising difficulty in being able to say what they mean by a surface in general. They know a particular surface when they see one, but how can we tell a computer or a blind man what it is about surfaces that they all have in common? Mathematicians began to be faced with this problem in the late nineteenth century as mathematics grew more complicated, and it took them many years to find a way of saying what they meant by a surface, that all mathematicians would understand. They had to be able to agree on what they were talking about before they could begin to *prove* things about surfaces, to do mathematics about surfaces. Their agreed statement about what a surface is, is called a *definition*; but it is not easy for beginners, so we shall approach the question in a different way. We shall eventually *make* a definition of our own, but in ordinary language that does not look mathematical. One can do mathematics in many dialects of a 'professional' language, and a mathematician chooses a particular dialect to suit his immediate purposes. Indeed, the following discussion is designed to show how mathematicians look at things they wish to study, and decide on the right words to use, in order to make their study easier.

Let us therefore ask a different question. How would you *make* a surface? (This might then help you to say what you mean by a surface: anything assembled according to your instructions would be a surface, although perhaps some exceptional 'surfaces' would not be made that way.) Now, most people only make surfaces as the 'skin' on some solid, by baking dough, moulding clay, or assembling wood or concrete forms. It is hard to say what we mean by a 'solid' and its 'skin' (has a jelly a surface; is it a solid?). But a seamstress makes a sort of surface when she stitches together pieces of cloth to make a dress, and an engineer makes a surface when he joins metal sheets to make the hull of a ship, the fuselage of a plane, or the body of a car. In all these cases, certain simple bits of surface – pieces of cloth, panels of metal – are being joined to make more complicated ones. We may not have the skill or tools of a seamstress or an engineer, but instead of pieces of cloth or metal we can use sheets of paper cut into polygons. These polygonal panels can be joined with sticky tape along edges, instead of being stitched or welded, to form more complicated things that most people will agree are surfaces. Some people might say that these paper surfaces are rather special, for various reasons, but let us consider the objections later, when we have considered such paper surfaces more closely.

1.2 TWO RULES FOR MAKING PAPER SURFACES

The simplest such paper surface, then, will be a single paper sheet, shaped into a polygon with at least three sides, but perhaps four, five, or more. Let us indicate the edges clearly by sticking tape (coloured black, say) along each one, as in Fig. 1.1; if we double the tape over the edge, the black stripe appears half-width. A polygon prepared in this way will be called a 'panel'.

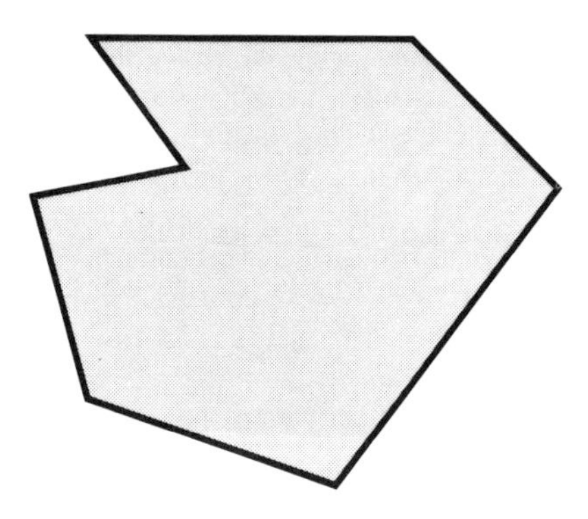

Fig. 1.1

Next, we can use the black tape to join two such panels along an edge, as in Fig. 1.2, and gradually build up constructions like those in Fig. 1.3. On each single panel, the tape appears half-width. Notice that in Fig. 1.2(*b*), we allow several edges of a new panel to be taped to edges of panels already assembled.

After some trials, we find we can bend our panels to make curved surfaces like that of the cylinder, or have three (or more!) panels with a common edge; see Fig. 1.3(*b*). Also we could use the black tape at a corner to make the constructions of Fig. 1.3(*c*), but this uses a different rule of construction from that shown in Fig. 1.2. To be quite clear about what we are allowing, let us lay down a rule of construction just as we would lay down rules if we were playing a game (we shall add another rule in a moment):

Rule 1
We can tape panels together only along their edges.

This rule, then, does not allow us to stick a panel by a corner, or to the middle of another. We could not, therefore, make the models shown in Fig. 1.3(*c*); and we cannot have Ⓣanything like *A*, *B*, *C* in Fig. 1.4(*a*). Instead we must get the same result as in Fig. 1.4(*a*), by following the order of Fig. 1.4(*b*). It will simplify things later if we take Rule 1 to mean that we *cannot* tape together two edges of the same panel.

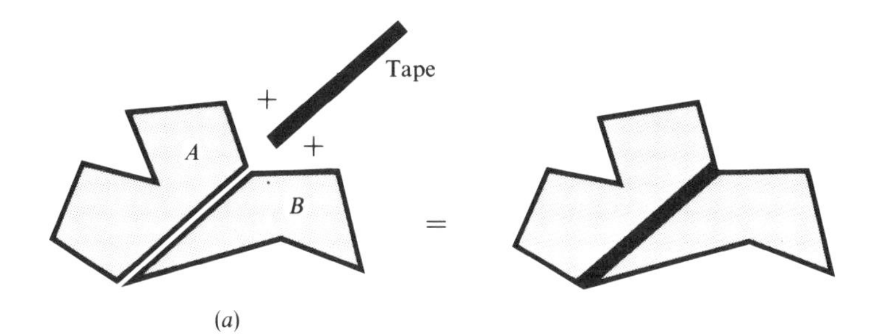

(a)

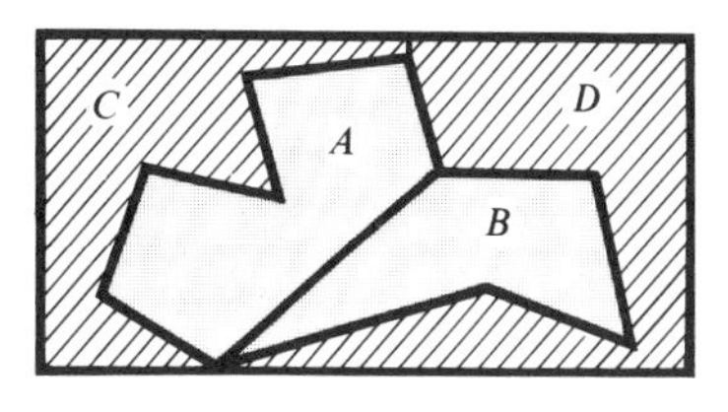

C has several edges joined to $A + B$
D has several edges joined to $A + B + C$

(b)

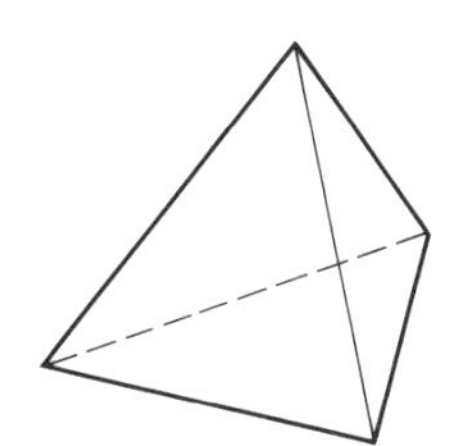

Tetrahedron

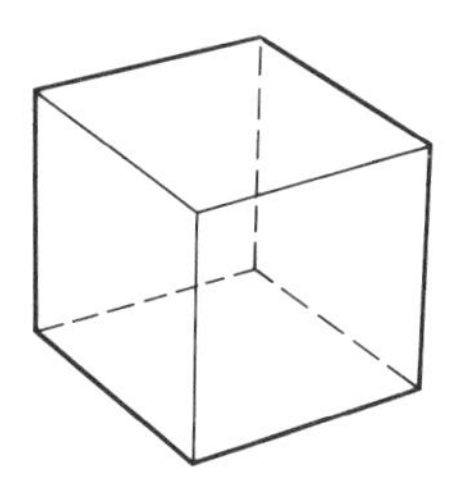

Cube

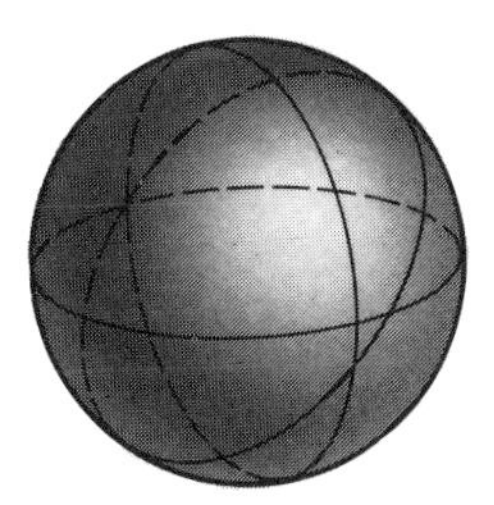

Sphere with panels

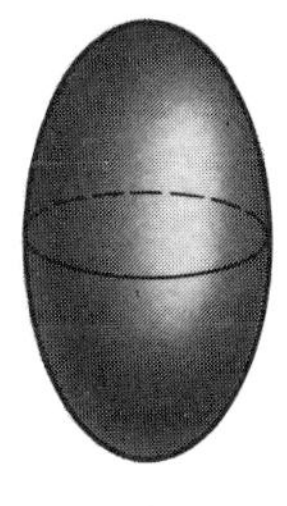

Ovaloid

(c)

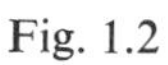

Fig. 1.2

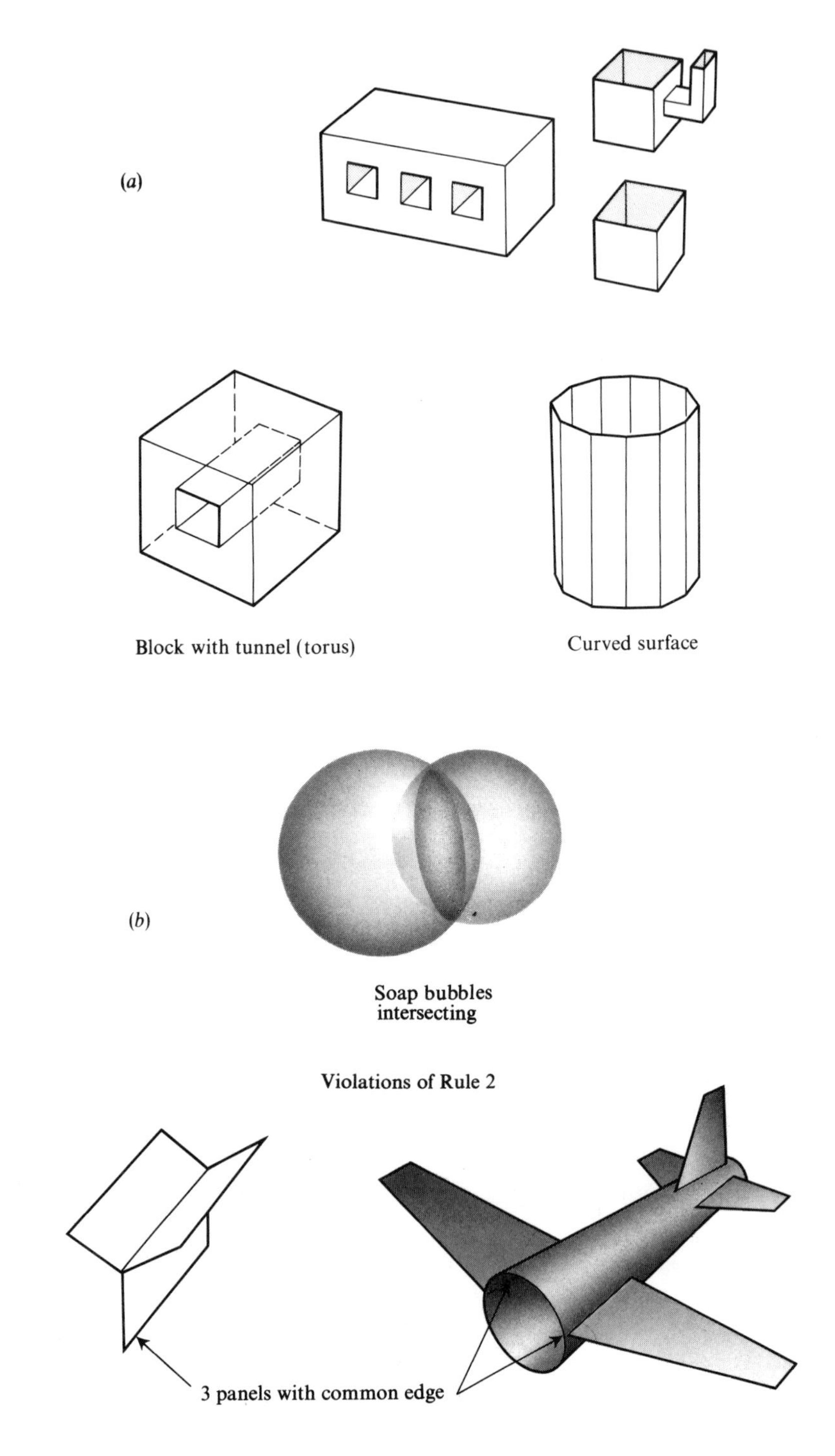

Violations of Rule 2

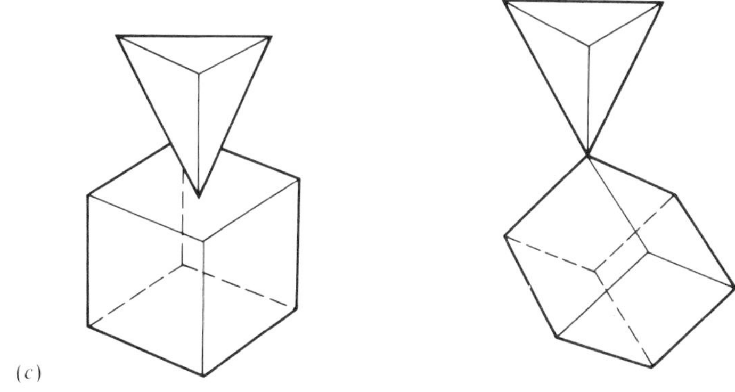

(c)

Violations of Rule 1

Torus

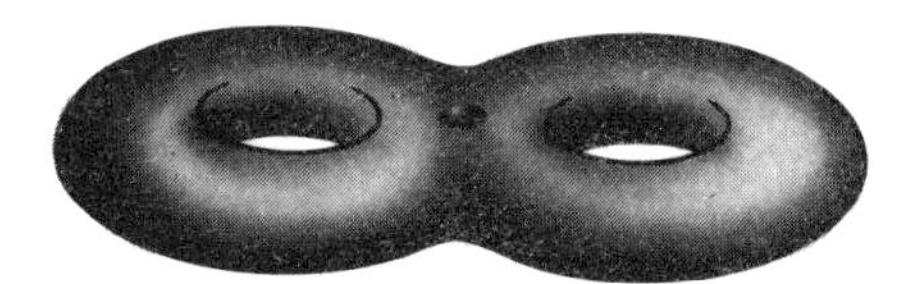

Double torus
(pretzel) with hole
(or puncture)

(d)

Surface of genus 4

Fig. 1.3

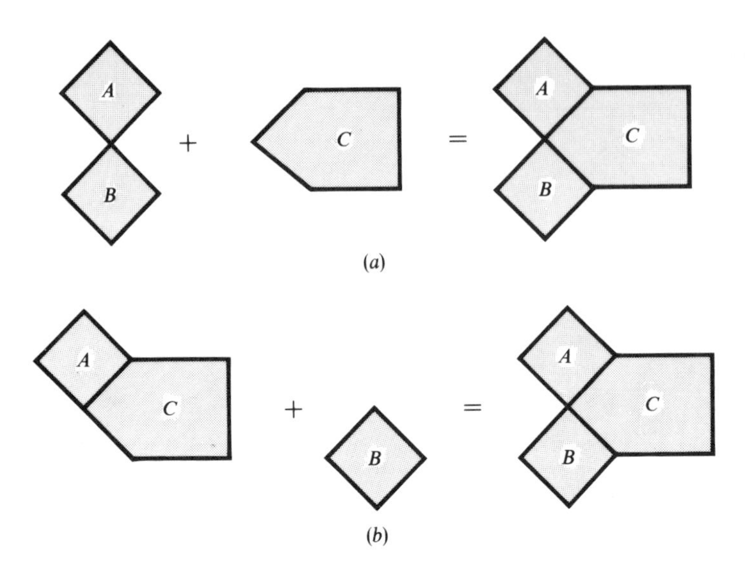

Fig. 1.4. (*a*) Wrong order; panels *A* and *B* were not taped along an edge. (*b*) Right order; an edge taped each time.

Many people will agree that these assemblies are good examples of surfaces, but some might make objections like this:

First Objection. 'These paper surfaces are rough, and the black tape sticks up against your hand when you stroke the surface.'

To which we could reply that by being sufficiently careful, and using sufficiently flimsy tape, we could make the panels as smooth as we like.

Second Objection. 'These paper surfaces have sharp corners and angles, but most surfaces we see are smooth like spheres or glass jars.'

To which we could reply that if we used enough *very tiny* panels, we could smooth out these surfaces as much as we liked, so that they would feel quite smooth. (In fact, many 'rounded' surfaces around us are made of millions of tiny crystals, as in the frontispiece, which have flat triangular or square faces.) So, *by using enough panels*, we can model most surfaces rather well by these constructions – in particular, those of Fig. 1.3(*d*).

Indeed, we could now model such a great variety of surfaces with these panels, that we shall need to exclude certain kinds, for simplicity. The ones where three or more panels are joined by an edge could be quite complicated, like the soap bubble in Fig. 1.3; but that kind of complication is not what we want to talk about at the moment. We therefore leave them out by agreeing to impose a second rule:

Rule 2

Any edge of the paper surface can belong to either one or two panels, but not to three or more.

Thus, once two panels have been joined by tape along an edge, that edge cannot be joined to any further panel. This rule excludes surfaces like that of Fig. 1.3(*b*), but our two Rules, 1 and 2, still allow us to make a very wide range of surfaces. We now want to describe these, and find out things about them.

Let us call Ⓣ[a] any assembly of panels, taped together according to both Rules 1 and 2, a **'paper surface'**. It will consist of a finite number of panels, with *edges* and *corners* indicated by the black tape. (These edges and vertices form what is known as a 'linear graph', Ⓣ[b] whose nodes are the corners of the panels. We shall not need the theory of linear graphs in this book, but the interested reader may consult Ore [15].)

1.3 SOME NAMES FOR THINGS

When we make paper surfaces according to our two rules, it helps to have names for some special parts, and for some of our more common surfaces. We have already mentioned the *panels*, the *edges* and the *corners*. By Rule 2, an edge belongs to just one panel, or to just two. The first kind of edge is called **free**, and we have no special name for the other kind – we can call them 'not free' if we need to talk about them. If there are no free edges at all, we call the surface **closed**. If there are some free edges, however, they form the **boundary** of the surface. Let us look at some examples.

The single panel in Fig. 1.1 has all its edges free, and its boundary is indicated by the heavy lines. It is customary to give the names 'tetrahedron', 'cube', 'torus', and 'surface of genus 4' respectively to the surfaces thus labelled in Figs. 1.2 and 1.3; each is a closed surface. We shall later say more about the strange name 'genus', but it refers to the numbers of 'handles' of which there are 4 in the sketch. The second surface in Fig 1.3(*d*) has a boundary and is called a 'double torus with one hole'.

Exercise 1.3

1. What do you think we would mean by a 'triple torus with five holes'? Sketch a quintuple torus with no holes. Is it a surface of genus 5, would you say?

2. For each surface S of those illustrated in Figs. 1.2 and 1.3(*a*) count the numbers of corners, edges and panels in S. If these numbers are C, E and P respectively, find the number $C - E + P$ in each case. (This curious alternating sum is interesting, as we shall see later.) Mark in panels on Fig. 1.3(*d*) and calculate similarly.

It might happen that the panels of a surface all have the same number of edges, while equal numbers of panels meet at each corner. The surface is then called a 'regular polyhedron'. For example, the tetrahedron and cube are regular polyhedra. More examples of regular polyhedra are shown in Fig. 1.5, and they look especially beautiful if all their panels are exactly the same shape,* with all edges and angles equal. (Their boundaries are then called 'regular polygons'.) You can understand the names better if you know that the ancient Greeks studied the regular polyhedra and called them the 'Platonic solids' after the famous philosopher Plato (fourth century B.C.) who thought about them a lot. (We shall avoid the word 'solid' here.) The Greeks called the panels 'hedra', so 'tetrahedron' means

* If these are merely quadrilaterals, the surface is called a hexahedron or cuboid, rather than a cube.

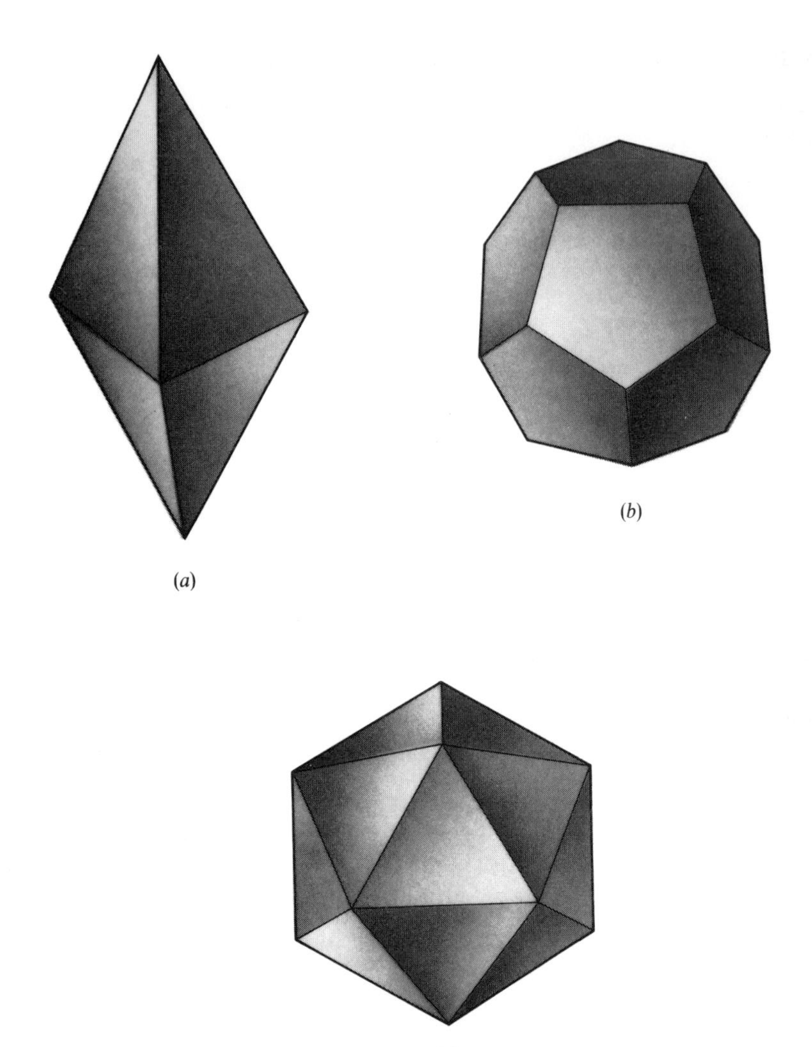

Fig. 1.5. The Platonic polyhedra other than cube and tetrahedron: (*a*) octahedron, (*b*) dodecahedron, (*c*) icosahedron.

'4-face', and the plural is 'tetrahedra'. The other names should now make sense, if we remember that 'octa' means 'eight', 'hexa' means 'six', 'icosa' means 'twenty', and 'dodeca' means 'twelve' (= 'two-ten'). These Platonic polyhedra are all *closed*, because they have no free edges. An interesting question asked by the Greeks was this:

'Are there any other kinds of regular polyhedron?'

and they were able to answer 'No', at least with a certain other restriction. We shall look at their explanation later (see Exercise 2.6, No. 13).

A very simple surface, which is not closed, is the **cylinder** in Fig. 1.3(*b*). It has some very close relatives, for example, the **prisms** of Fig. 1.6; and another

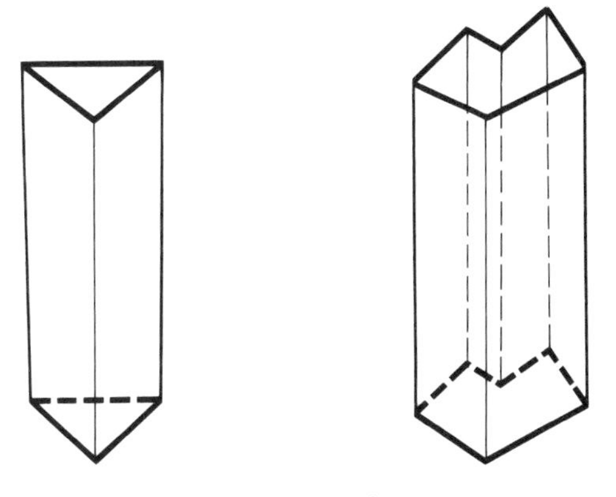

Fig. 1.6. Prisms.

surface in the same family is the **annulus**, which is like a disc with a hole punched in it. ('Annulus' is a Latin word meaning 'ring', and most people think of an annulus as circular, as in Fig. 1.7(*a*); but we use the name for any shape that looks like a panel with a single hole, even if it is made with

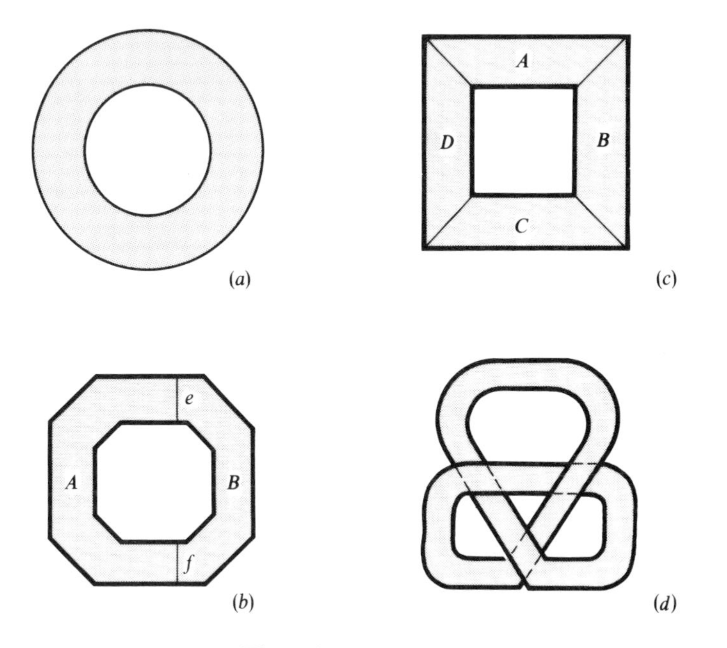

Fig. 1.7. Annuli.

several panels like those in Fig. 1.7(*b*) and (*c*).) The cylinder, the prisms and the annuli all have free edges, and so *they are not closed surfaces*. Each has a boundary, consisting of two endless chains of free edges. Each such chain forms a loop which does not cross itself. We call such loops **Jordan curves** after the French mathematician C. Jordan (1838–1922) who first studied them. If a surface has one or more boundary curves, we call it a **surface with boundary**; thus a surface is either closed, or else it is a surface with boundary.

Let us look again at the annulus in Fig. 1.7(*b*). It is very simple, because it has only the two panels *A* and *B*, taped together along the edges *e* and *f*. Now, if we asked a friend to do the taping, he could give us a surprise we didn't expect! For, after taping along the edge *e*, he might give the panel *B* a twist before he taped the edge *f*. The result would look like one of the pair shown in Fig. 1.8,* and it is called a **Moebius band** (or Moebius **strip**) after the German mathematician A. F. Moebius (1790–1868) who first wrote about it. Notice that it is not flat like the annulus and has only *one* Jordan curve for its boundary. The two types of band in Fig. 1.8 arise because our

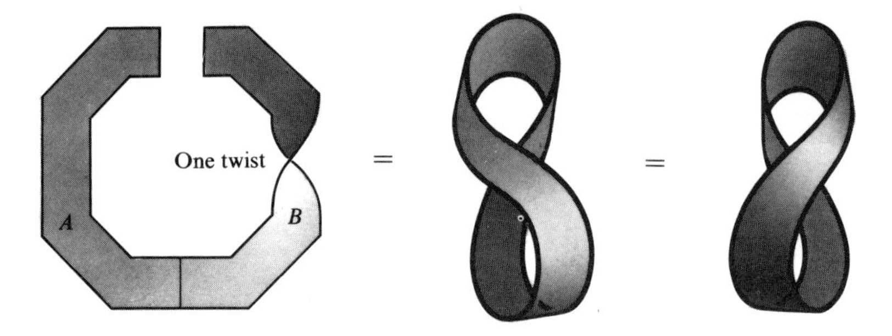

Fig. 1.8. Moebius band.

friend could have given the panel *B* a left-hand twist, or a right-hand twist. If we reflect either of the two bands in a mirror, the image will be the other band, so they resemble each other in the way our left hand resembles our right. Because of this, we shall not mind which direction of twist is given.

You can check that this kind of effect would occur also if the panel *B* had been twisted not 1 but 3, 5, or any *odd* number of times. If *B* has been twisted 2, 4, or any *even* number of times, the result would be a surface with two boundary curves, just as with the annulus in Fig. 1.7(*b*), where *B* was twisted no times (still an even number of times!). Usually we shall only think of a panel with no twist or one twist, and as we said above, we ignore Ⓣ the direction of the twist.

1.5 THINKING ABOUT A MOEBIUS BAND

The Moebius band has some strange properties when we compare it with the annulus, and these properties are sometimes shown off by conjurors.

* The reader is not expected to draw for himself pictures of this quality each time we study a Moebius band. We shall later use an easier and more schematic type of drawing, but each time an unfamiliar surface is introduced, we use a realistic picture for clarity.

For example, if we use scissors to cut an annulus along its length, it falls apart into two thinner annuli. What happens if we make such a cut along the length of a Moebius band? To work it out, remember that we showed earlier that the Moebius band was really constructed from a rectangle (formed when the two panels A and B were taped along the edge e). In Fig. 1.9, we have shown only a rectangle R, with the edge e in the centre

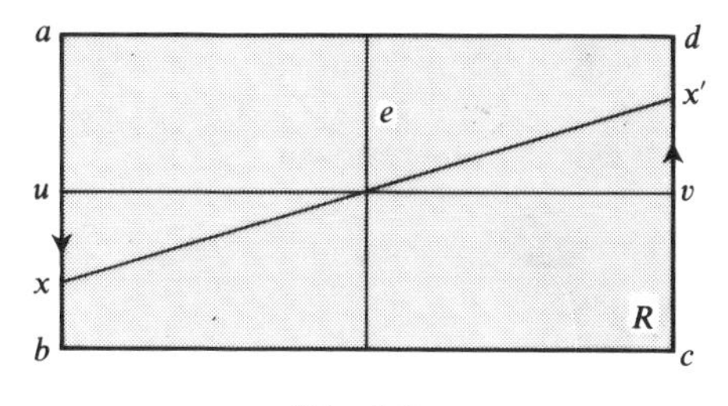

Fig. 1.9

for simplicity. To complete the taping to form a Moebius band, we have to join each point x on the edge ab to the point x' on the opposite edge cd (see Fig. 1.9). But we have to twist the edge cd once before taping, and this is the same as letting x' run *up* cd as x runs *down* ab. (If instead, both x and x' run downwards, we would simply get an annulus again.)

Now let us think what happens if we cut the Moebius band along a line down its middle. We would obtain the same effect if we first cut the rectangle R along the central line uv in Fig. 1.9, and then taped as before. But we can see from Fig. 1.10 what would happen: the edges ub and vd get taped together (giving a rectangle with *one* twist), and then the edges au and cv are taped together to form a continuous band, and this involves a second twist. Thus we get a 2-twisted band, which has two edges, $ad + bc$ and $u'v' + u''v''$, this last formed from the two edges of the cut.

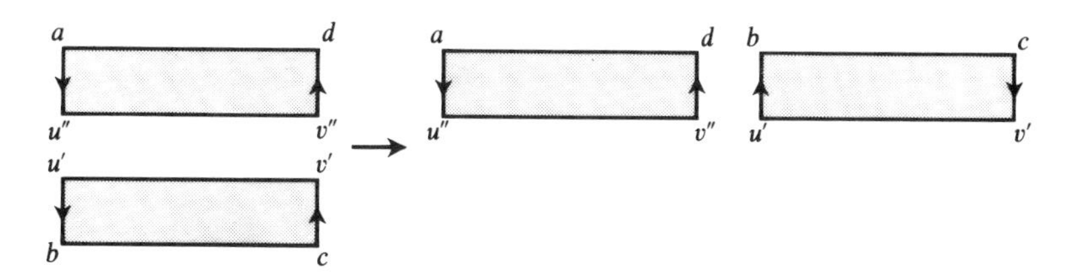

Fig. 1.10. Slicing a Moebius band down the middle.

Exercise 1.5

1. Show that if we cut along the Moebius band, keeping always one-third of the width from the boundary, then the strip falls into two circular loops, one of which is a 2-twist band with 2 edges, the other a Moebius band whose width is one-third of what we started with. [Hint: Think of the cut as being along the lines pq, $q'p'$ shown in Fig. 1.11.]

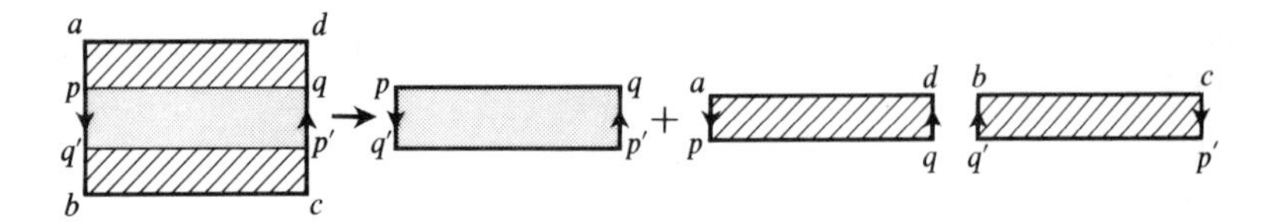

Fig. 1.11. Slicing a Moebius band near its edge.

2. Work out how one band links the other in the last problem. N.B. Perform some experiments with scissors and paste. You may then feel that the diagrams are less complicated than the practical experiments!

1.6 WHERE HAVE WE GOT TO?

Let us now tidy up our procedures a little. We began by trying to agree about what a surface is. We decided to restrict ourselves to paper surfaces, and we made Rules 1 and 2. Thus, we agreed to say that we shall call anything a paper surface if it results from an assembly of panels, taped together one after the other, according to Rules 1 and 2.

We then looked at the simplest surfaces that could be made in this way. In the next chapter we shall look at more complicated ones.

Exercise 1.6

1. A person shows you a paper surface he has made, but does not tell you the order in which he assembled the panels. Do you think you could work out a possible order of assembly? (For a discussion, see Chapter 7, p. 75.)

The remaining questions of this set are for those readers familiar with coordinate geometry.

***2.** Make sketches of the sets of points with coordinates (x, y, z) that satisfy the equations

(*a*) $x^2 = 1$,
(*b*) $x^2 + y^2 = 1$,
(*c*) $x^2 + y^2 + z^2 = 1$,
(*d*) $x^2 + y^2 - z^2 = 0$, 1 and -1 (3 surfaces).

Which of these 'surfaces' do you think could be modelled by paper surfaces? (See Appendix C, p. 105.)

***3.** Describe the graph of a torus, in (x, y, z)-coordinates. Is your description as useful as if you had modelled the torus as a paper surface?

***4.** What surfaces are represented by the equations

(*a*) $z^2 = r^2 - a^2$, (*b*) $z^2 = b^2 - r^2$, (*c*) $z^2 = (1 - r^2)(r^2 - c^2)$,

where $r^2 = x^2 + y^2$ and (in (*c*)) $c^2 < 1$. [Hint: First find where $z = 0$; always confine yourself to those (x,y) which give a positive right-hand side, since $z^2 \geqslant 0$.]

***5.** Show that the equation (in which $r^2 = x^2 + y^2$):

$$z^2 = (1 - r^2)[(x - a)^2 + y^2 - u^2][(x - b)^2 + y^2 - v^2],$$

where $0 < a < a + u < b - v < b < b + v < 1$, represents a closed surface of genus 2. Construct similar equations for surfaces of genus 3, 4, etc.

***6.** A line segment of length 2 passes at its mid-point M through the point $(2\cos u, 2\sin u, 0)$ in (x, y, z)-coordinates, and it makes an angle $u/2$ with the upward vertical through M. Show that as u varies from 0 to 2π, the line segment sweeps out a Moebius band, and the typical point on it has coordinates

$$x = 2\cos u + v\sin\frac{u}{2}\cdot\cos u,$$

$$y = 2\sin u + v\sin\frac{u}{2}\cdot\sin u,$$

$$z = v\cos\frac{u}{2},$$

where $0 \leqslant u \leqslant 2\pi$, $-1 \leqslant v \leqslant 1$.

2. Making complicated surfaces

2.1 PLANAR REGIONS

The annulus in Fig. 1.7(*a*) is flat, while the Moebius band is not. In fact, we could cut an annulus in one piece from a flat sheet of paper, and then add tape to indicate panels. By the same method, we could cut out a flat sheet with several holes, and mark out panels on it afterwards. Of course, by Rule 1 (p. 2) we could not mark out the panels in just any old way. For example, we could not have the parts P, Q in Fig. 2.1 added to the

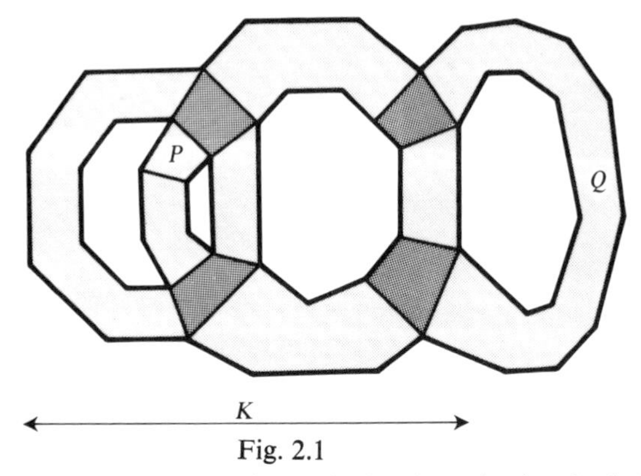

Fig. 2.1

piece K unless edges had first been formed like those in the shaded panels. (Incidentally, Rule 1 does not prohibit us from taping more than one edge of a new panel to a paper surface.) Paper surfaces of this kind are called **planar regions** because they form regions ('parts') of plane sheets. Notice that the boundary contains one Jordan curve[T] for each hole, together with one more Jordan curve, called the **outer boundary**. (We first met Jordan curves on p. 10.)

2.2 ADDING AN EAR

Once we have a planar region, we can make a new hole in two ways. One way is to tear a puncture in it, but this method does not use the two rules we agreed on above. The other way is to add, to the old surface, a new panel (like Q in Fig. 2.1) by taping two of its edges to the *same* Jordan curve in the old boundary. A name is suggested by the picture,* so, for short, we

* We shall often draw boundaries without indicating the corners in them, and, for simplicity, we shall not always draw all the panels in a surface. The reader is strongly advised to construct paper models of all the surfaces we mention, to add greater understanding to his study of our sketches. It suffices (and is quicker) to staple the ears and to add the 'black tape' later, by drawing it with a fibre tip.

shall say that we have **added an ear** to the old surface. Notice that the two edges that we use in the taping must not be consecutive edges of the new panel Q, otherwise we would not have a hole. Thus, a panel cannot be used as an ear unless it has at least *four* edges.

Notice that when we add an ear, we increase by one the number of Jordan curves in the boundary of the planar surface. This does not happen

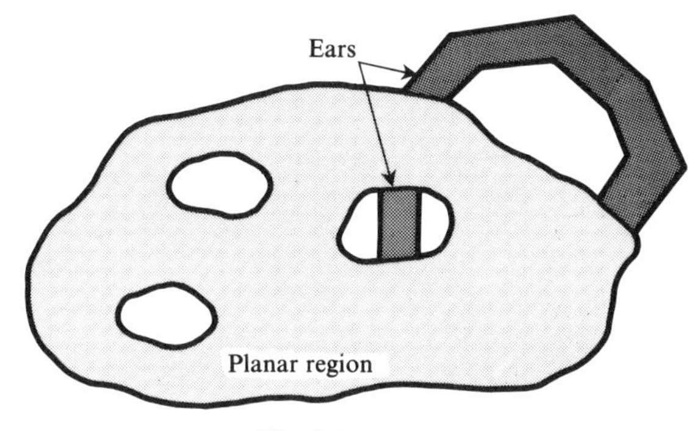

Fig. 2.2

with our own ears on our heads, because they are solid! Of course, we can go on adding new ears to a surface, and need not stop after one or two. We could also add a new ear in such a manner that it threaded through the existing holes, to make under- and over-passes (as in Fig. 1.7(d)), or looped itself several times round an old ear. (A really florid example is shown on the cover of this book.) In our sketches, however, we shall not illustrate such complications because they will make no difference ⓣ to our deductions.

Exercise 2.2

1. If we add 5 ears to a panel, how many Jordan curves are there in the boundary of the resulting paper surface?

2. If a planar region has h holes, how many boundary curves has it?

2.3 NON-PLANAR SURFACES

We have just seen that we can make a planar region more complicated by adding an ear; this gives us one extra hole. But before the second of the

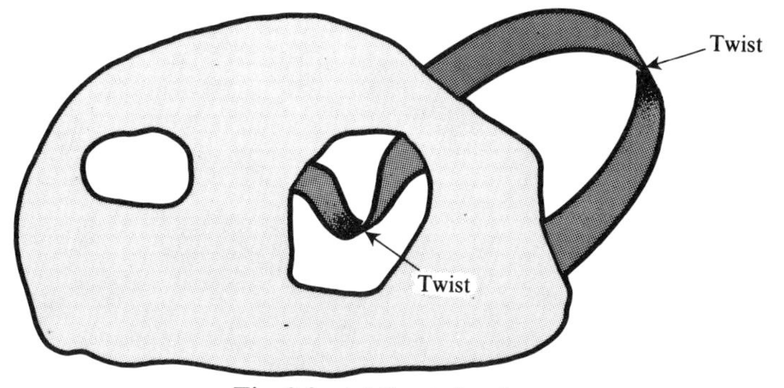

Fig. 2.3. Adding twisted ears.

two edges was taped to complete the ear, we could give the panel a twist (just as with the Moebius band) and then tape it. The result is a union of a plane region and a Moebius band, as in Fig. 2.3. It cannot be flattened out to be planar, as we shall show later on. We shall call this method of adding a new panel **adding a twisted ear**. Obviously we can add an ear, or a twisted ear, to any paper surface, not only to planar ones, provided the panels match the ends of the ears, like the shaded ones in Fig. 2.1. For example, if we add a twisted ear to a planar region, we can then add an ear (untwisted or not) to the result although it is non-planar.

Exercise 2.3

1. How many extra boundary curves are created when we add a twisted ear?

2. 7 ears are added to a panel, of which 3 are twisted. Show that the resulting paper surface has exactly 5 boundary curves.

3. Show that if we add an ear to a panel, then we get an annulus, and if we add a twisted ear to a panel we get a Moebius band.

4. Let S be a paper surface and P a twisted ear. What happens if we cut along the centre-line of P, and continue the cut on to S to make a complete round-trip? (See Exercise 1.5, No. 1.) If $S+P$ were planar, such a cut would separate $S+P$ into two pieces. Hence conclude that $S+P$ is *not* planar!

2.4 RECOGNIZING A TWIST

A word of caution is now necessary, concerning the difference between a twisted, and an untwisted, ear. When we add these to planar regions, they are easily recognized, but when they are added to more complicated surfaces, we need a more certain way of recognizing them.

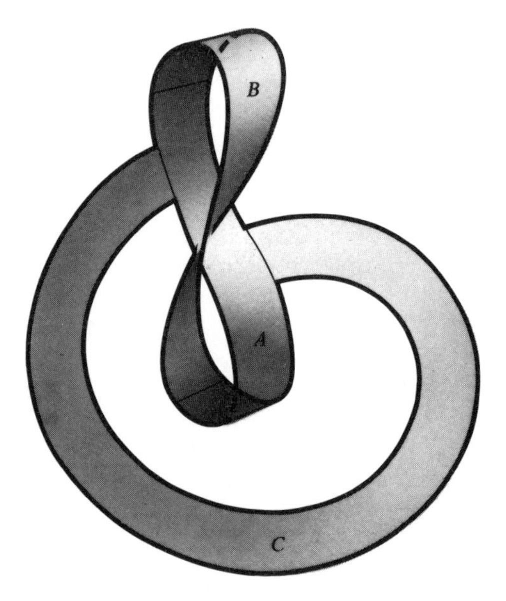

Fig. 2.4

Consider the example shown in Fig. 2.4, where the panel C appears to be added as an ear to the Moebius band M (formed from panels A and B). However, the resulting paper surface S has only *one* boundary curve! If, using a paper model, we slide one end of C round the boundary of M, we find that it starts to look twisted as we move it far from the starting point. But *we* think it looks twisted then, because we can refer it to fixed objects around us, in our 3-dimensional world. To a short-sighted beetle, who lived on the surface and could not see any further than the edge, C would seem as 'flat' as before we moved it. We shall meet other examples later, where such twisting can mislead us, so we need a different kind of test.

However, boundary curves can be counted, by reference only to the surface and not to the outside world. Even our short-sighted beetle could do that (provided he could count at all). In future, therefore, we use the following test for deciding whether or not an ear is twisted:

Test

Let S be any paper surface, planar or not, and suppose P is a panel taped to S as an ear. If S and $S+P$ have the same number of boundary curves, then P will be called **twisted**. *If they do not, then P is* '**untwisted**'.

This test forces us then to call C in Fig. 2.4 a twisted ear. An ear P either does or does not pass the test for being twisted, so here 'not twisted' is the same as 'untwisted'. It follows from the test that, in the case when P is untwisted, $S+P$ has one more boundary curve than S. Since C in Fig. 2.4 *appears* untwisted to anyone who stuck it on, the question of its direction of twist is not meaningful here.

2.5 ADDING A BRIDGE

When we add an ear to a paper surface, we tape the ends of the panel to the *same* Jordan curve in the boundary. Why shouldn't we tape the ends to two *different* Jordan curves in the boundary, as in Fig. 2.5? A panel stuck on in this way will be called a **bridge**. If we had twisted it once, before taping the second end, then we would call it a **twisted bridge**. Adding

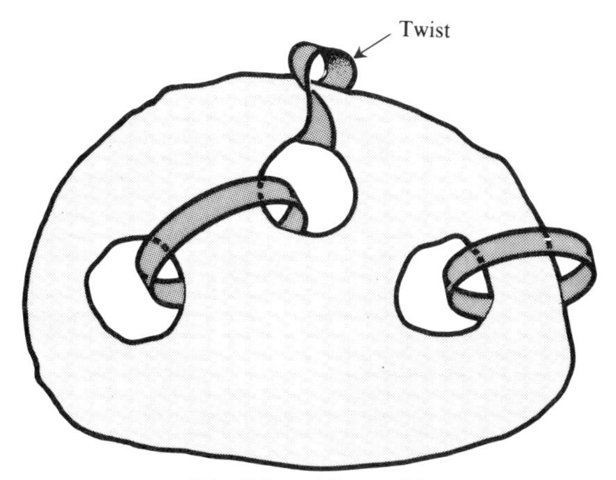

Fig. 2.5. Adding bridges.

bridges always gives us non-planar surfaces. As with ears, a panel cannot be used as a bridge unless it has at least *four* edges.

An example of a twisted bridge is also given in Fig. 2.4, where the panel B can be thought of as a twisted bridge that joins the two boundary curves of the annulus $A + C$. It will turn out that to add a twisted bridge amounts to adding two twisted ears (see Fig. 4.22), after losing a hole. See also Remark 2 on p. 31.

Exercise 2.5

1. Show that if we add a bridge to a paper surface S, then the resulting surface has 1 boundary curve fewer than S, whether the bridge is twisted or not.

2. Let us add 8 ears to a panel, of which 3 are twisted; and then add 5 bridges to the result, of which 2 are twisted. Show that the resulting paper surface has exactly 1 boundary curve.

3. Why can a bridge not be added to a Moebius band or to a single panel?

4. What is the maximum number of bridges, twisted or not, that can be added to a panel with 8 ears?

2.6 THE EULER NUMBER

As we add more panels as ears or bridges, we build ever more complicated surfaces. To keep a check on the amount of complexity, we can count various features like the number of boundary curves, the number of holes, etc. For this purpose we need some mathematical notation. Suppose we have made a paper surface S, and then we find that its boundary consists of a certain number of boundary curves. We denote this number by* β, or β_S to show that it depends on S. For example

$$\beta_S = 1, \text{ if } S \text{ is a panel,}$$
$$\beta_S = 2, \text{ if } S \text{ is an annulus or cylinder,}$$
$$\beta_S = 1, \text{ if } S \text{ is a Moebius band.}$$

The great Swiss mathematician, L. Euler (1707–83) found that it was interesting to look at a number which we shall denote by* χ or χ_S, the **Euler number** of S. For this, we first count the numbers of corners, edges and panels in S and let

C denote the number of corners of S,
E denote the number of edges of S,
P denote the number of panels of S.

Then, to work out the number χ_S we add C to P, and subtract E to get

$$\chi_S = C - E + P.$$

For example, if S is a single panel then it has just 1 boundary curve, and exactly as many corners as edges, so $C = E$ while $P = 1$; therefore† $\beta(\text{panel}) = 1$, $\chi(\text{panel}) = C - C + 1 = 1$. For the tetrahedron in Fig. 1.2, $C = 4 = P$ and $E = 6$ so $\chi = 4 - 6 + 4$, i.e. $\chi = 2$. Since the tetrahedron is a closed surface, $\beta = 0$.

* β and χ are the Greek letters beta and chi respectively; and χ is used because the Euler number is often called the 'characteristic'.

† We read 'β (panel)' as 'β of a panel'. Similarly when β is changed to χ, or 'panel' to some other word.

For some paper surfaces S, χ_S may well be negative if E is bigger than $C+P$. See for example Exercise 1.3, No. 2, where the punctured double torus had $\chi=-3$; and Fig. 2.4 where $\chi=-1$.

Exercise 2.6

1. Let two panels be taped along an edge to form a paper surface S. Show that $\beta_S=1$ and $\chi_S=1$.

2. If a surface S has $\chi=-3$, find possible numbers of ears and bridges that might be added to a panel to form S.

3. Let S denote a paper surface, and let T denote the result of adding one untwisted ear to S. Show that the numbers β_S, β_T, of boundary curves satisfy

$$\beta_T=\beta_S+1,$$

while for the numbers χ_S, χ_T (even if the new ear were twisted)

$$\chi_T=\chi_S-1.$$

4. With S as above, let U denote the result of adding one bridge (twisted or not) to S. Show that

$$\beta_U=\beta_S-1, \qquad \chi_U=\chi_S-1.$$

***5.** If you had difficulty with No. 2 above, try it after Nos. 3 and 4.

6. If 21 ears and 13 bridges are added to a single panel, to form a surface S, calculate β_S and χ_S.

***7.** Let S and T be two paper surfaces, each with boundary, and suppose they are joined by taping along 1 edge of each, to form a paper surface V. Show that the numbers C, E, P, β and χ are related by the equations

$$C_V=C_S+C_T-2, \quad E_V=E_S+E_T-1, \quad P_V=P_S+P_T,$$
$$\beta_V=\beta_S+\beta_T-1, \quad \chi_V=\chi_S+\chi_T-1.$$

8. Suppose that S is a closed surface, and that after the assembly of its panels, there are recorded u ears and v bridges (twisted or not). Show that $\chi_S=2-u-v$.

***9.** S is a closed surface, and for each corner v we let $M(v)$ denote the number of edges that meet in v. Show that $2E=M(v_1)+M(v_2)+\cdots+M(v_C)$, where v_1, $v_2, \ldots, v_C$ are the corners of S. Hence show that the number of corners v, for which $M(v)$ is odd, is even. Also show that $lC\leqslant 2E\leqslant CM$, when the numbers $M(v)$ lie between l and M; and hence that $2E\leqslant C(C-1)$. (No two edges have the same pair of end-points, by definition of a panel.)

***10.** If, in the last question, all the numbers $M(v)$ are even, find all the possibilities for them, if E is 7. Do any of these correspond to an actual paper surface?

11. S is a closed surface, and for each panel Q we let $m(Q)$ denote the number of edges of Q. Show that

$$2E=m(Q_1)+m(Q_2)+\cdots+m(Q_P),$$

where $Q_1, Q_2, \ldots, Q_P$ are the panels of S. [Hint: Imagine that S is 'exploded' into separate panels, and then observe that *two* edges of panels are joined up to make *one* of S.]

12. In No. 11, show that if $m(Q)$ is the same number q for all the panels, then $2E=Pq\geqslant 3P$.

***13.** Using the notation of No. 9, suppose that each number $M(v)$ is constant, say p. Suppose also that $\chi_S = 2$. Show that

$$\frac{2E}{p} - E + \frac{2E}{q} = 2, \quad \text{so that} \quad \frac{1}{p} + \frac{1}{q} > \frac{1}{2}.$$

Now show that, if p and q are whole numbers $\geqslant 3$, the only possible pairs satisfying this equality are $(p,q) = (3,4), (3,5), (5,3), (4,3), (3,3)$.

(This result tells us that there are no Platonic polyhedra other than those listed in Section 1.4, subject to the restriction that $\chi = 2$. (See also Ore [15].) What can you say for other values of χ, e.g. $\chi = 0$?)

2.7 THE PUNCTURED TORUS

The simplest surface to which we can add a bridge is an annulus. Let us look carefully at what we get, as the resulting surface is a very basic building block for other surfaces. Consider first the case when the bridge is *untwisted* (see Fig. 2.6(*a*)); call the resulting surface a **bridged annulus**. Since the bridge joins the two Jordan curves that form the boundary of the annulus, the new surface has only *one* boundary curve:

$$\beta(\text{annulus} + \text{bridge}) = 1.$$

We can check from the figure that $\chi(\text{annulus} + \text{bridge}) = 1$. In Fig. 2.6(*b*), we show a disc, curved over so that its boundary matches the bold curve

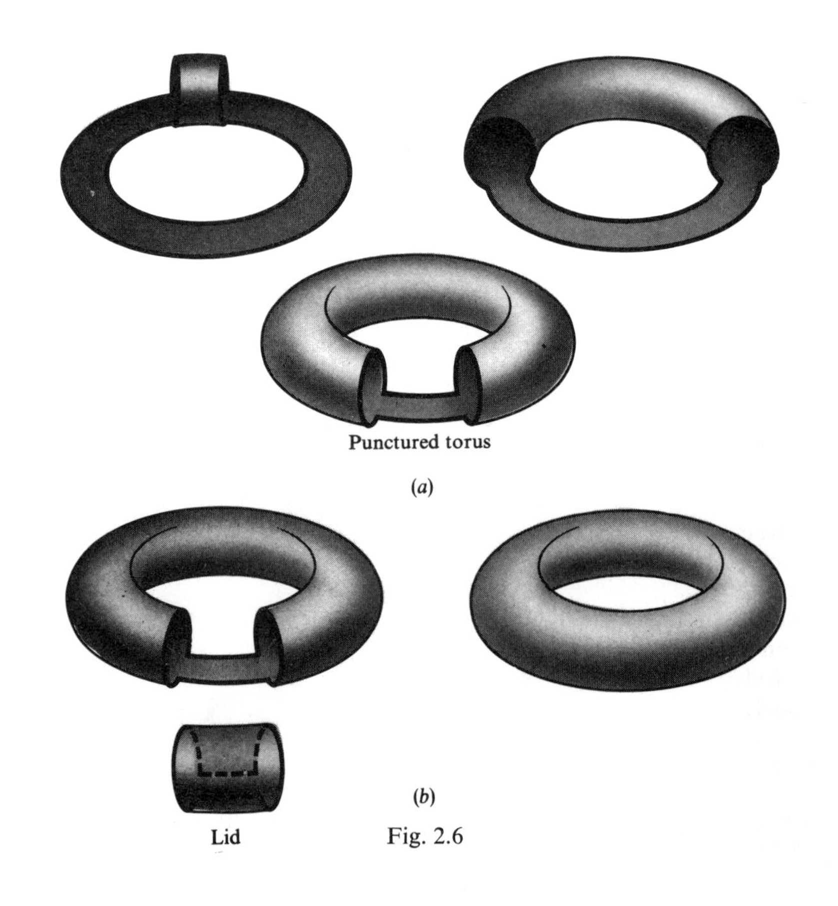

(*a*)

(*b*)

Fig. 2.6

in Fig. 2.6(*a*). This disc forms a sort of 'lid' for the bridged annulus, and if we fit it on, we get a torus. Thus

$$\text{bridged annulus} + \text{lid} = \text{torus},$$

so

$$\text{bridged annulus} = \text{torus minus lid}.$$

We usually call a torus minus a lid a 'punctured torus', because it is rather like a bicycle tyre without a patch. A more common name for a bridged annulus is therefore a **punctured torus**

2.8 THE KLEIN BOTTLE, AND THE PROJECTIVE PLANE

In our previous discussion of Fig. 2.4, we saw (p. 18) that the surface there could be thought of as the result of adding a *twisted* bridge B to an annulus. We saw too that we obtained a surface with just *one* boundary curve and with $\chi = -1$. Now, we were thinking of the surface in two different ways that we can write algebraically Ⓣ[a] as

$$\begin{aligned} K &= \text{annulus} + \text{twisted bridge} \\ &= (A + C) + B = B + (A + C) \\ &= (B + A) + C \\ &= \text{Moebius band} + \text{twisted ear}. \end{aligned}$$

Thus, Fig. 2.7 is another representation* of K. As with the punctured torus, we could try to fit a lid to K, but the fitting cannot be done in our 3-dimensional world since Ⓣ[b] the lid gets in the way of the rest of the surface. This

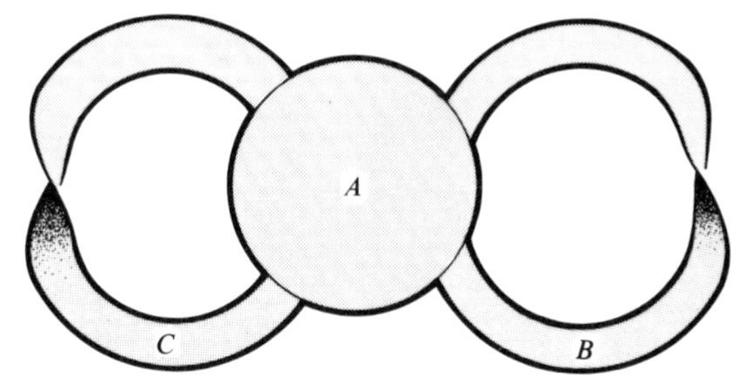

Fig. 2.7. Moebius band and twisted ear.

obstruction need not prevent our discussing the result, however, which is a closed surface called a **Klein bottle**, after the German mathematician Felix Klein (1849–1925) who first thought of it. It is a closed surface, whose Euler number is zero. Therefore the surface in Fig. 2.7 is called a punctured Klein bottle.

Models of Klein bottles are sometimes seen, like that in Fig. 2.8. These models have had to be allowed to have walls that cross themselves, to get them into our 3-dimensional world, rather as a 2-dimensional picture of the Moebius band shows its edge crossing itself; and Fig. 2.9 shows their

* Here is an example of a simplified form of sketch that we shall gradually develop, in order to help the reader to avoid the difficulties of making realistic drawings.

relationship to the Klein bottle formed above from panels. It shows what we would get if we sawed the glass bottle in half and laid out the two halves side by side. Each of these is a Moebius band of which two separate arcs of its boundary have been stuck together (in violation of one of our rules?). By an **arc**, we mean a chain of edges that neither crosses itself, nor joins up to form a Jordan curve.

A related matter concerns that of adding a lid to the boundary of a Moebius band. Again we cannot do it in our 3-dimensional space, but the resulting closed surface can be discussed, and mathematicians call it a **real projective plane** –

$$\text{Moebius band} + \text{lid} = \text{real projective plane.}$$

(The curious name occurs because of the branch of Geometry called Projective Geometry, in which the real projective plane was first discovered. See Griffiths and Hilton [7], Chapter 17; 'real' refers to the 'real numbers' as opposed to the 'complex numbers'. Thus a Moebius band is a punctured real projective plane.)

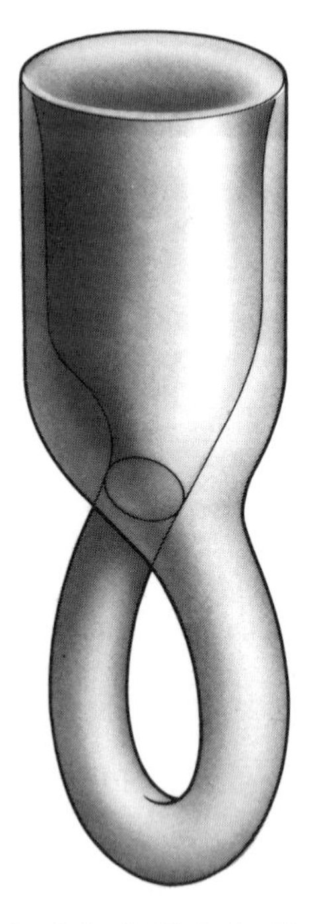

Fig. 2.8. A Klein bottle.

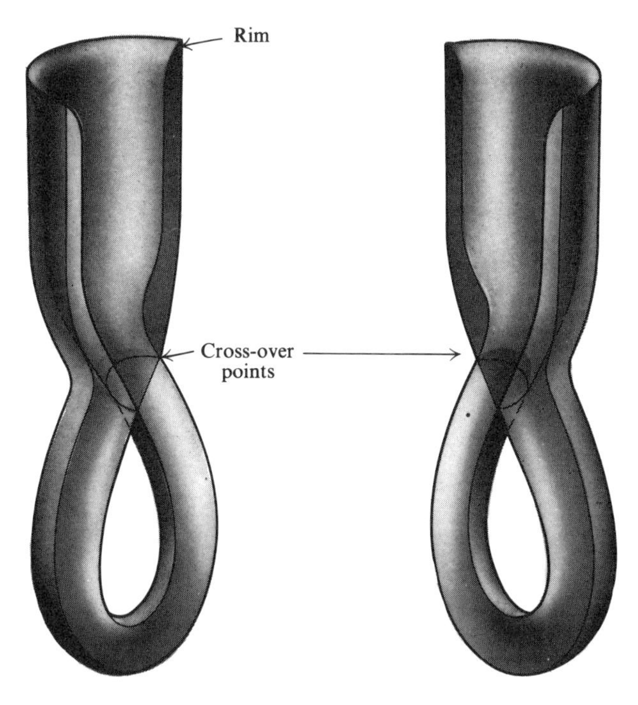

Fig. 2.9. Cleaving a Klein bottle into two halves. The rim of the Klein bottle is not involved in the boundary of either half. The cross-over point occurs in this model because it is built in our own space of three dimensions. These points need not occur if we had four dimensions to work in.

Exercise 2.9

Calculate the Euler numbers of the Klein bottle and the real projective plane, using the panellings described above.

Remark. The ends of panels B and C in Fig. 2.4 alternate round the boundary of panel A, whereas in Fig. 2.7 they do not. This might cause the reader to question our assertion (p. 21) that Fig. 2.7 is another representation of Fig. 2.4. Our justification is given with Fig. 4.10 below, and involves a discussion of how we agree to slide an end of an ear or a bridge around a boundary curve.

3. Multifarious complications

3.1 BUILDING UP AND BREAKING DOWN

Once we have built a supply of simple things, like annuli, Moebius bands, punctured toruses and punctured Klein bottles, we can start joining these together to make more complicated paper surfaces. On the other hand, if someone gives us a strange new surface, we might be able to study it by splitting it into the simple ones we know, and by seeing how they fit together. First let us look at an example of this 'breaking-down' process, in connection with ears.

3.2 MULTI-BRIDGES AND MULTI-EARS

When we add a bridge or an ear to a surface S, we tape *two* edges of the panel P that forms the bridge or ear. But there is nothing in Rules 1 or 2 to stop us from taping even more edges of P, as in Fig. 3.1, such that *either* they all lie on one old boundary curve (so we might for the moment call P a 'multi-ear') *or* distinct arcs of P lie on distinct boundary curves of S (so we might call P a 'multi-bridge') or a mixture of the two (when we call P a 'multi-ear and bridge'). Twists can obviously be made *before* taping as well. The number of taped arcs of P is the *multiplicity* of P; an ordinary bridge or ear then has multiplicity 2. .

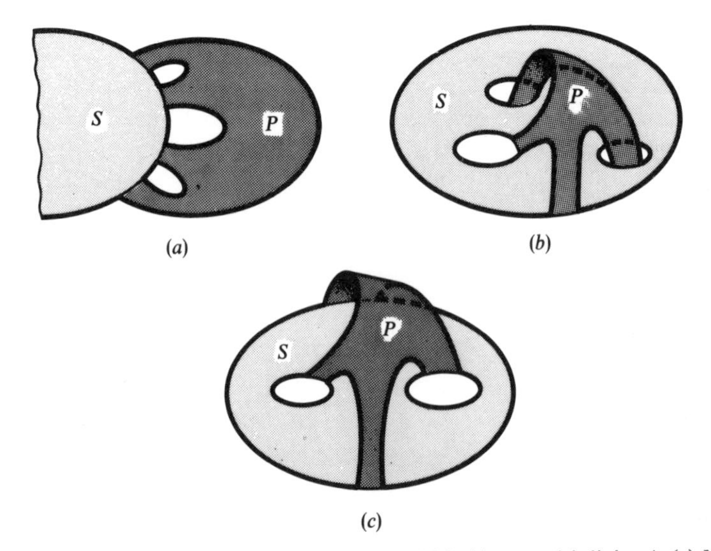

Fig. 3.1. (*a*) Multi-ear: multiplicity 4. (*b*) Multi-bridge: multiplicity 4. (*c*) Multi-ear and bridge: multiplicity 4.

This construction gives us nothing new however. For, let us divide P into two panels Q and R as indicated by the dotted line l in Fig. 3.2, then $S + P = (S + Q) + R$ and Q is an ordinary bridge or ear added to S (perhaps twisted). Also, R is a multi-ear or multi-bridge, or multi-ear and bridge added to $S + Q$ but *less complicated than P was*, because the multiplicity of R is 1 less than that of P. We can now split off another bridge or ear from R, and repeat the process until we have replaced P by h ordinary bridges and ears, where $h + 1$ is the multiplicity of P. The only difference is that we have had to add tape along the dividing lines that we

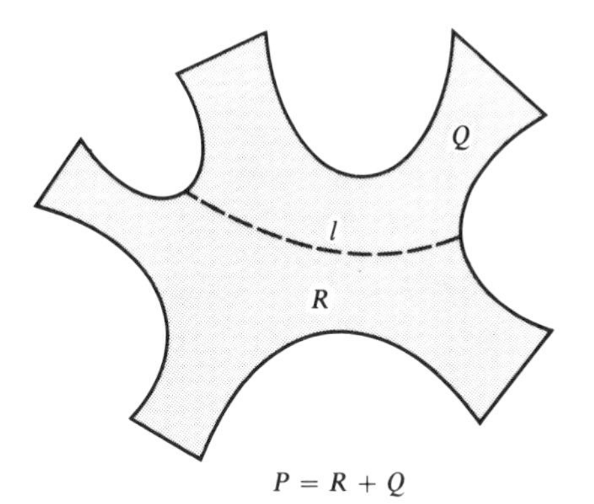

$P = R + Q$

Fig. 3.2. Multiplicities of P and R are 5 and 4.

introduced (like l), so $S + P$ has been slightly 'repanelled' to get it in the form

$$S + P_1 + \cdots + P_h,$$

where the P's are ordinary bridges or ears, perhaps twisted.

We may therefore confine ourselves to the use of ordinary bridges and ears for building paper surfaces, because they will yield for us everything that multi-bridges and multi-ears could do.

Exercise 3.2A

Investigate the changes in the boundary number β and the Euler number χ that result from the addition of a multi-bridge or ear.

We have seen how a panel P may be taped to a paper surface S, to fill in a hole, like a lid. If we do not seal it all the way round, but tape only some of its edges consecutively – for if we skipped some, we would have a multi-ear or multi-bridge – then the result is rather like taping along a single edge. We then say that we have taped P along an *arc-succession* of edges (because the tape follows a curve called an *arc*, which has two ends and does not cross itself). This method of taping is very useful, and we use a special symbol for it, writing

$$S \dot{+} P$$

(adding a dot above the plus sign) to indicate that P is added neither as a lid, nor as a multi-ear, nor as a multi-bridge. If P is added as a lid, we write

$$S \overset{\circ}{+} P.$$

Exercise 3.2B

Show that, in Fig. 1.2, the tetrahedron T can be written (don't be afraid of the brackets!)Ⓣ:

$$T = [(P \dotplus Q) \dotplus R] \mathring{\dotplus} V,$$

while the cube C can be written

$$C = [([(P_1 \dotplus P_2) \dotplus P_3] \dotplus P_4) \dotplus P_5] \mathring{\dotplus} P_6.$$

Find similar expressions for the other regular polyhedra.

3.3 THE PUNCTURED SPHERE WITH g HANDLES

If we take several punctured toruses, we can join them together to make a more complicated surface, as follows. As in Fig. 3.3, suppose we had, say, four punctured toruses, each perhaps in the form of a bridged annulus (see Section 2.7). We could then tape each one of them to an octagonal panel P as shown in Fig. 3.3(*a*), where we see that we get a surface with just the one boundary curve shown. In Fig. 2.6, we saw that a bridged annulus was a punctured torus with a large puncture, and if we had made the puncture smaller we would have got a picture like that in Fig. 3.3(*b*). Comparing with Fig. 1.3, this surface is now like the surface of genus 4 there, but with a large puncture; and if we make the hole a bit smaller we get Fig. 3.3(*c*), a **punctured sphere of genus 4**. Obviously, we could have used any number, say g, of punctured toruses to start with, provided we had used a panel with $2g$ sides instead of the octagon in Fig. 3.3(*a*). The number g has a special name: it is called the **genus** of the surface. Clearly the punctured

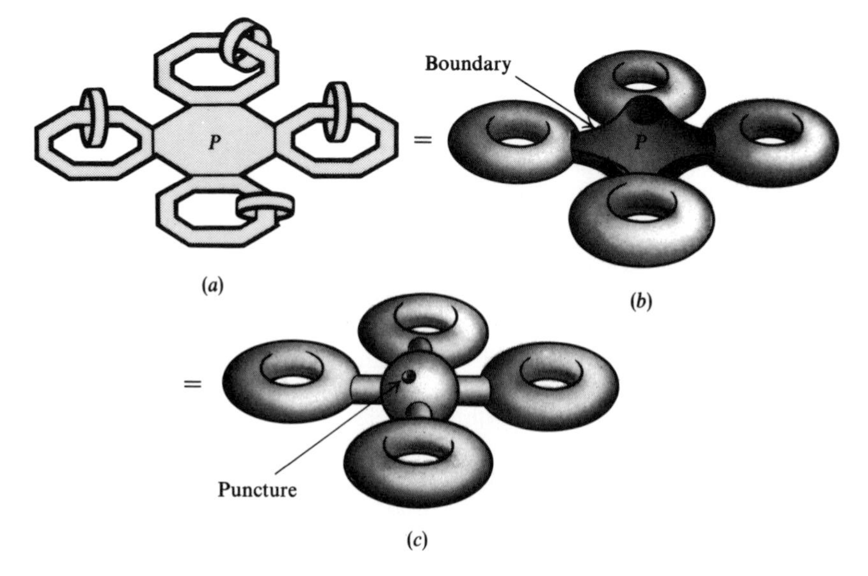

Fig. 3.3. Punctured surface of genus 4.

torus is a sphere with 1 handle, provided we do not mind in that case about the 'blister' that the panel P forms on the side.

Exercise 3.3

1. Calculate the Euler number of the punctured sphere of genus 4. (Use the result of Exercise 2.6, No. 7.) What happens to this number if we 'repair the puncture', by adding a lid, to get a closed surface?

2. Show that the punctured sphere of genus g has Euler number equal to $1 - 2g$.

Instead of using punctured toruses, we could have used some punctured Klein bottles (of the sort we met in Section 2.8), but then the resulting surface does not look like anything very familiar. We could also add ears (twisted or not) to either kind of surface, but the resulting paper surfaces have no special names. See, however, p. 44 below.

3.4 THE CLOSED SPHERE WITH g HANDLES

We have seen that if S is a punctured sphere with g handles, then S has a single boundary curve, consisting of a polygon with (say) h sides $e_1, \ldots, e_h$. This curve surrounds a 'hole' in the paper surface; and we can fill in the hole with a lid by taping a single panel Q of h sides $f_1, \ldots, f_h$, so that e_1 is taped to f_1, e_2 is taped to f_2, and so on. The panel Q may possibly need to be bent a bit to get it to fit, although by reorganizing the other panels we could get a neat fit as in Fig. 3.4. Then $S \overset{\circ}{+} Q$ is a *closed* paper surface, and it is called a **sphere with** g **handles**. As before, we call g the **genus**. When g is 1 we get a torus, and when $g = 2$ we get a double torus. When $g = 0$ we simply get a (not very round) sphere, formed from the starting panel P (see Fig. 3.3) and the panel Q, which are joined like the upper and lower hemispheres of the Earth along the Equator.

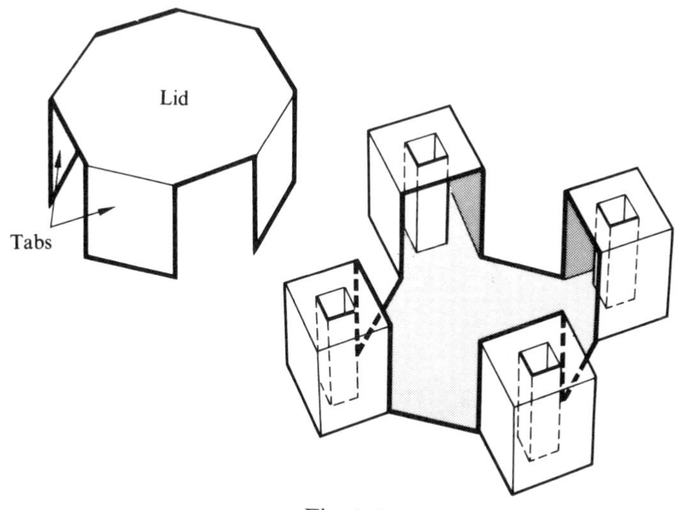

Fig. 3.4

Exercise 3.4

1. Calculate χ (torus) and χ (double torus).

2. Let S be a paper surface, assembled from panels $P_1, P_2, \ldots, P_m$, and suppose that u of these are added as ears, and v are added as bridges (twisted or not). If none of the panels is added as a lid, show that $\chi_S = 1 - u - v$. If $u = a + b$, where just b of the ears are twisted, show that $\beta_S = 1 + a - v$. (See Exercise 2.6, Nos. 3 and 4.)

3. If, in addition, w of the panels are added as lids, show that $\chi_S = 1 + w - u - v$, and $\beta_S = 1 - w + a - v$. Hence show that if S is a closed paper surface, then $a = w + v - 1$ and $\chi_S = 2 - 2v - b$. (Thus $\chi_S \leqslant 2$.) Show further, that if $\chi_S = 2$, then no ear is twisted, there are no bridges, and $a = u = w - 1$; while if $\chi_S = 1$, then $b = 1$, $v = 0$ and $u = w$.

3.5 ORIENTING A SURFACE

There is a certain simple, and yet subtle, property that some paper surfaces have, which others do not. Suppose, for example, that we take a fairly simple subdivision, into panels, of an annulus or punctured torus; then we can draw 'orienting' arrows, circulating round the panels, as shown in Fig. 3.5(a). The arrows have two basic properties,[1] one 'internal', the other 'external'.

Internal Property
If e is an edge of two panels, their orienting arrows pass along e in opposite directions.

To discuss the external property, suppose that an edge e lies in just one panel (and hence in one boundary curve J of S). The orienting arrow for the panel then flows from one end of e to the other, giving us a *direction* of e. In this way, all the edges in J acquire a direction. Then the arrows in Fig. 3.5(a) satisfy the following 'external' property.

External Property
The directions, of all edges in any boundary curve of S, are never opposed.

If we look in Fig. 3.5(b) at the orienting arrows, on the punctured Klein bottle, then we see that the Internal Property fails on the edge uv (as a

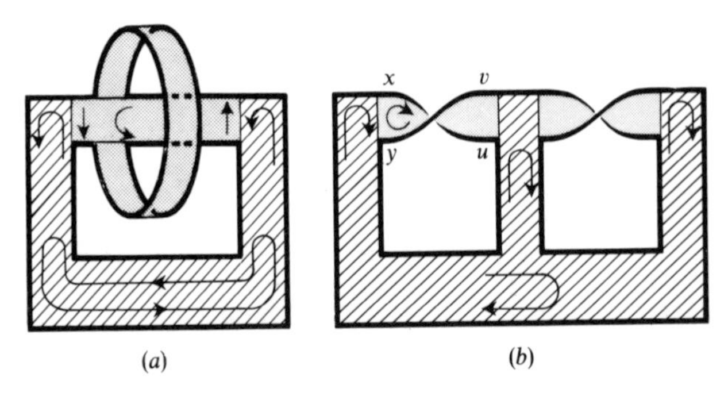

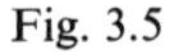
Fig. 3.5

result of making it hold on the edge xy). Also the External Property fails, for example, because the direction of xu is opposed to that of yu.

We can try to add orienting arrows to the panels of any paper surface S, in such a way that both the Internal and External Properties are satisfied. If we can manage it,Ⓣ then S is said to be **orientable** and we have **oriented** S; if it is impossible, then S is **non-orientable.**

If S is orientable and has boundary curves, then each time we close one of these with a lid, we can always add to the lid, an orienting arrow that circulates in the *opposite* direction to the arrows on the boundary (opposite, in order to satisfy the Internal Property). The resulting closed surface is then orientable also. Closed orientable paper surfaces are 'two-sided' because they enclose a volume that we cannot enter without piercing the

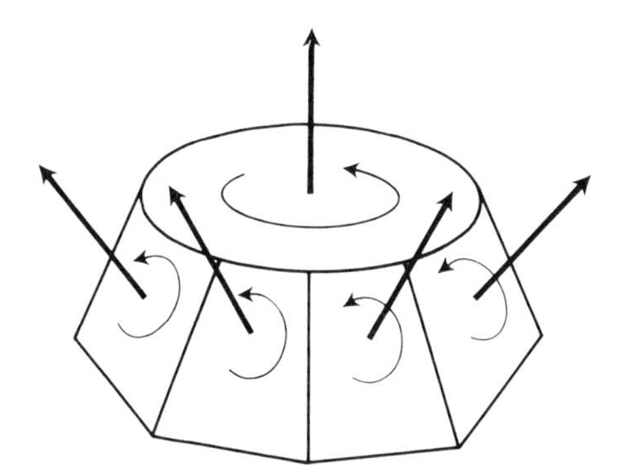

Fig. 3.6. Orienting a lid.

surface. (By contrast, although a glass model of the Klein bottle (see Fig. 2.9) encloses sufficient volume to store water for keeping flowers in it, we can get at the water without penetrating the glass – indeed to top up the water.)

To see why an orientable closed surface is 2-sided in the above sense, we use the orienting arrows to assign a new arrow perpendicular to each panel P, giving it the direction in which a screw would move if we twisted it with a screw-driver in the direction of P's orienting arrow. These new arrows then all point 'inwards' or all point 'outwards'.

Exercise 3.5

1. Divide up into panels, a punctured sphere with 2 handles, and show that the resulting panelled surface is orientable.

2. Suppose that the punctured toruses in Fig. 3.3 are all divided into panels in such a manner that each is orientable (but they may be divided in different ways from each other). Show that the resulting sphere with 4 handles is orientable.

3. Show that the Platonic polyhedra (Fig. 1.5), and the closed sphere with two handles, are orientable. Check that they are two-sided, using the screw-driver rule mentioned above.

4. Show that if a surface has been oriented, and we then reverse the directions of all the arrows, we get another orientation of S.

5. If a closed surface S has an inside and an outside in our ordinary physical space, show how to use arrows pointing outwards, together with the screw-driver rule, to orient S. Show that a non-orientable closed paper surface cannot be built so as to have an inside and outside.

6. A paper surface S is orientable, and a new panel P is added. Show that $S+P$ is orientable, if P is added as an untwisted ear or untwisted bridge. Show also that $S \dotplus P$ is orientable. What happens if we allow twisted ears or bridges?

3.6 IMPORTANT QUESTIONS

In this chapter, we have seen something of the rich profusion in the 'zoo' of paper surfaces. Now, when we go round an animal zoo, we often wonder how to recognize certain animals – 'which families are they in?'. We might also wonder how comprehensive the zoo is – 'Are these all the snakes there are?'. Similar questions arise with our surfaces.

For example, the closed surface $S+Q$ in Section 3.4 is like the surface of a solid block of wood, and the solids we see in daily life all have surfaces that are modelled by the spheres with g handles (for some g). But, you may object, in what sense does the surface of genus 4 in Fig. 1.3 'model' that of Fig. 3.7? (The two surfaces each have four tunnels, but they are *arranged*

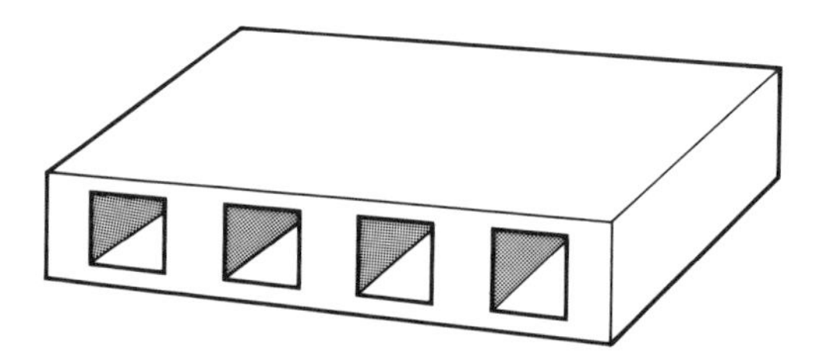

Fig. 3.7. Block with 4 tunnels.

differently.) We must now discuss this important question, and begin a new chapter. However, you will probably first ask a related but even bigger question:

The Big Question
Are there any other paper surfaces, made according to Rules 1 and 2, of a kind really different from the ones described so far?

Before we can give an answer, we must ask what this question means, because everything depends on what we mean by 'different'. Then Ⓣ we shall show that we can answer 'No' to the question, but it will need a lot of work to make sure that we have not overlooked anything. This work of making sure that we know what the question means, and that we are agreed on the answer, is called a **mathematical proof.**

Exercise 3.6

1. Can the surface of a pair of spectacles (without lenses) be modelled by a closed paper surface of the sort constructed above? Of what genus? What happens if we include the lenses?

2. What is the genus of the surface of (i) a kettle, (ii) a saucepan, (iii) a pair of scissors? (Regard them all as being made of *thick* metal.) Draw plans to model their surfaces using paper surfaces.

3. What is the genus of the surface of (i) a saucer, (ii) a cup, (iii) a kitchen table, (iv) a chair, (v) a spoon, (vi) a fork, (vii) an egg-whisk, (viii) a salt-cellar, (ix) a pepper-pot?

Remark 1. These exercises are intended to support the assertion (p. 30) that the solids we see in daily life all have surfaces *modelled by* spheres with handles. That assertion is related to the meaning of 'model' and of the word 'different' in the Big Question; and the exercises may help the reader to begin the thinking we develop later. Some practice is needed in distinguishing between a solid and its surface, trying to think of the surface alone as if the solid had been dissolved away. Because such a surface encloses a volume of solid, then it is *orientable* by the argument of p. 29. This is why non-orientable surfaces are less familiar to us.

Remark 2 (*twisted bridges*). At this point we can say more about how to identify a twisted bridge since visual inspection can be highly misleading. Let S be an oriented surface with orienting arrows satisfying the conditions of p. 28. Let P be a new panel whose edges u, v are taped to different boundary curves U, V, of S; it forms a bridge but we wish to decide whether or not P is twisted. By the External Property (p. 28) the arrows of S all go round U in the same direction, and similarly for V. Therefore, we can attempt to orient P to make the Internal Property hold on $S+P$. If such an orientation exists for P then we call P an untwisted bridge and $S+P$ is orientable; if not, then P is a twisted bridge. This process will not work if S is non-orientable, for then we have no consistent way of relating a direction round U to one round V. It turns out not to matter, as we see on p. 56.

4. Families of surfaces

4.1 LAMINAS AND COCOONS

When we have previously discussed the annulus and the Moebius band, we have thought of them as being formed from just two panels. Now in Fig. 1.7(*c*), an annulus is illustrated that has 4 panels, and we can easily divide it up into more – just as we can think of a Moebius band composed of many panels. Nor do they need to be particularly neat – we would still think of the paper surfaces illustrated in Fig. 4.1 as having[T] 'family

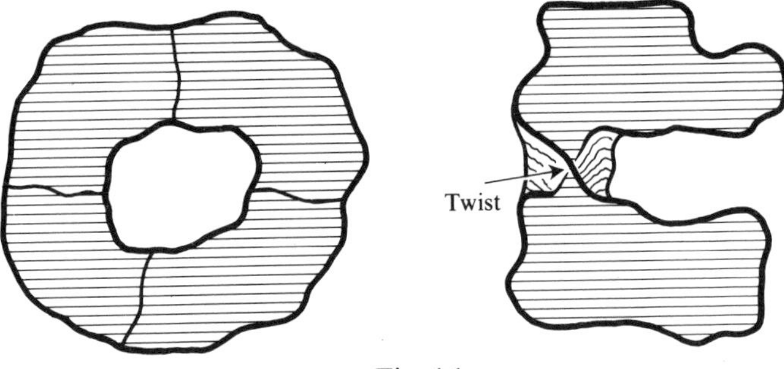

Fig. 4.1

resemblances' to the annulus and Moebius band respectively, even if we find it hard to express this resemblance in words. Similarly, referring back to Fig. 1.4, if we tape the two panels A and C together there, the result seems to have a 'family resemblance' to a single panel, and the resemblance persists when we tape the third panel B. Even more, suppose that we add a panel D to a paper surface S with boundary, by taping along an *arc-succession* of the edges of D, as explained in Section 3.2, so that as there we can write $S \dotplus D$ to indicate that we have neither used up *all* the edges of D (i.e. D is not a lid) nor skipped edges (i.e. D is neither an ear nor a bridge).

Using the notation $S \dotplus D$, we are agreeing that when D is a panel, then $S \dotplus D$ and S are both in the same family of paper surfaces: more briefly we write the following agreement:

Agreement 1

$$\text{Family}\,(S \dotplus D) = \text{Family}\,(S),$$

that is to say, the family of $S \dotplus D$ is the same as that of S. In particular,

$$\text{Family (panel} \dotplus \text{panel)} = \text{Family (panel)}.$$

Let us use the old-fashioned name **lamina** for anything in Family (panel).

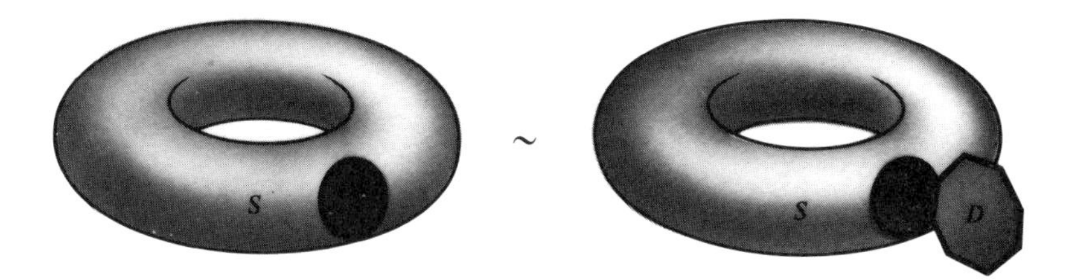

Fig. 4.2. Agreement 1 illustrated (the symbol ~ indicates that the surfaces are in the same family).

Thus the set or family that we call Family (panel) contains any paper surface of the forms

$$\text{panel} \dotplus \text{panel}, \qquad \text{lamina} \dotplus \text{panel}, \qquad \text{lamina} \dotplus \text{lamina},$$

and we call each of these three forms a lamina. Hence we arrive at:

Agreement 2

$$\text{Family (lamina} \dotplus \text{lamina)} = \text{Family (panel)}.$$

Further, the successive building-up of a lamina by adding a panel at a time shows that we may get ups and downs in the resulting paper surface, but

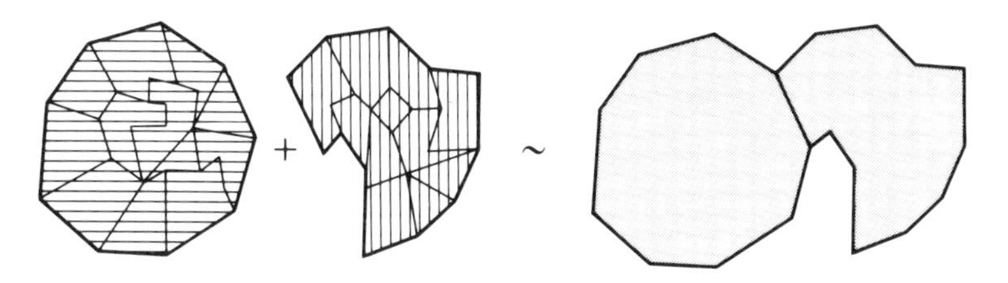

Fig. 4.3. Agreement 2 illustrated.

no bridges or holes. A lamina may contain cavities, like the punctured box in Fig. 1.3(a), or extensions as with a capped chimney. In general, then, a lamina may at worst look like a rocky, hilly countryside.

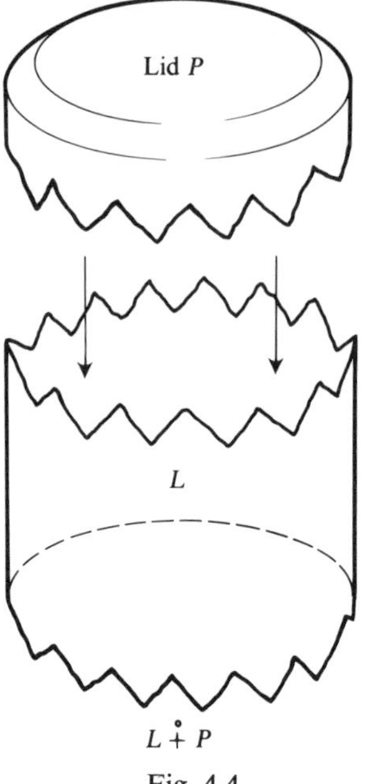

Fig. 4.4

A lamina has a single boundary curve, which we may then wish to close by adding a lid. Regardless of which shape of lamina or which shape of lid we choose (provided their edges fitted), we would expect the family resemblance to persist. Thus we shall suppose that we can record:*

Agreement 3

$$\text{Family } (L \mathring{+} P) = \text{Family } (M \mathring{+} Q),$$

whenever L and M are laminas, while P and Q are lids. Any paper surface in this family will be called aⓉ **cocoon**, so we are making a *definition* (i.e. an agreement always to use a term in a fixed way).

Definition

$$\text{Family (lamina } \mathring{+} \text{ lid)} = \text{Family (cocoon).}$$

The most basic example of a cocoon is any spherical or ovaloid surface, see Fig. 1.2. As another example, recall from Chapter 1 the Platonic polyhedra. We shall now make an assertion and *prove* it. The assertion and the fact that it can be proved, together form what mathematicians call a *theorem*:

Theorem

The Platonic polyhedra are all in the cocoon family.

Proof. In Exercise 3.2B, we saw that we could express the tetrahedron T as

$$T = [(P \dotplus Q) \dotplus R] \mathring{+} V.$$

* The equation is hard to state briefly in words, which is precisely why we have introduced the symbols!

Now $P \dotplus Q$ is a lamina L, as we agreed above in Agreement 1, and hence $L \dotplus R$ is a lamina M. Hence $T = M \overset{\circ}{+} V$ so, by the definition above, T is in Family (cocoon).

For a cube C, we obtained the expression

$$C = [([(P_1 \dotplus P_2) \dotplus P_3] \dotplus P_4) \dotplus P_5] \overset{\circ}{+} P_6$$

in Exercise 3.2B. As with T, we gradually get rid of the brackets to obtain $C = L \overset{\circ}{+} P_6$, where L is the lamina $[([(P_1 \dotplus P_2) \dotplus P_3] \dotplus P_4) \dotplus P_5]$. Thus C is of the form (lamina $\overset{\circ}{+}$ lid), so C also lies in Family (cocoon).

The remaining Platonic polyhedra can be dealt with similarly, where we regard one panel as the lid, and the rest as a lamina. This completes the proof of the theorem.

A person might say that this theorem does not impress him because, for example, a cube is less complicated than an icosahedron. In reply we must stress that we are here trying to ignore all the 'lesser' differences between surfaces, to concentrate on their most basic similarities. We are thinking here that the most basic property of a cocoon is that it can be constructed according to the formula lamina $\overset{\circ}{+}$ lid. (Remember: to a zoologist, apes and men belong to the same family, even though an anthropologist may study differences between families of men.) We hope our procedures will seem more satisfying, as the reader lets them soak in, and gains experience of surfaces through doing exercises.

Exercise 4.1

1. A rectangular dining table has square legs. Show that an exact paper replica of its surface is in the coccon family. Similarly for a brick, and a drawer without a keyhole.

2. If L and M are laminas, show that $\chi(L \dotplus M) = \chi(L) + \chi(M) - 1$.

3. Deduce Ⓣ from No. 2 above that $\chi(\text{lamina}) = 1$.

4. Show that $\chi(\text{cocoon}) = 2$.

4.2 THE ANNULUS FAMILY AND ITS RELATIONS

Again, the basic feature in our description of an annulus on p. 9 is, that it is built from two laminas L, M which are joined according to the prescription

$$L +_e M, \tag{1}$$

where '$+_e$' signifies that M is added as an ear.* (For convenience of reference, we shall number such prescriptions and equations.) For example, the four panels A, B, C, D in Fig. 1.7(c) are assembled in this way, since the annulus illustrated there is

$$A + B + C + D = (A \dotplus B) +_e (C \dotplus D),$$

and $A \dotplus B$, $C \dotplus D$ are laminas by our Agreement 2. As another example of how we recognize that something is in the annulus family, consider Fig. 4.5(a), in which we have added a 'chimney' to the hole in a square

* This formula would need modification if we had interpreted Rule 1 (p. 2) as allowing the opposite edges of a rectangular panel to be taped together.

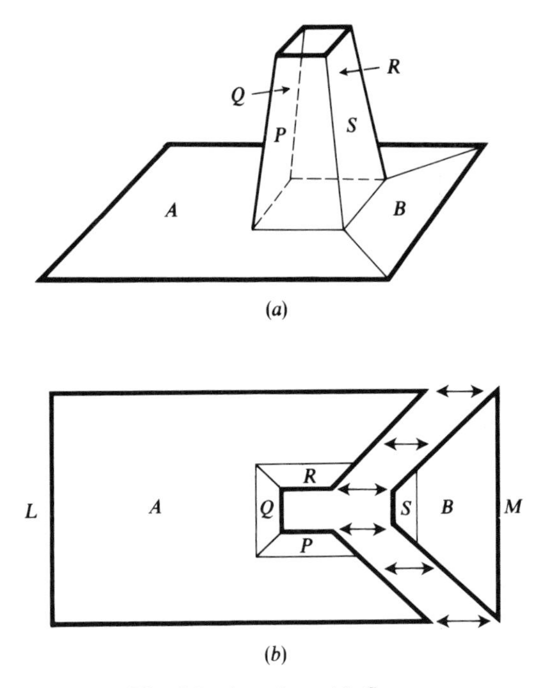

Fig. 4.5. Annulus with flange.

annulus. The exploded view in Fig. 4.5(*b*) shows that the surface is of the form $L +_e M$, where L and M are the laminas

$$L = ((A \dotplus Q) \dotplus R) \dotplus P,$$
$$M = B \dotplus S,$$

and the heavy arcs show where the ear M is joined to L.

For brevity, we sayⓉ that 'Fig. 4.5(*a*) is an annulus' when we really mean that 'the figure represents a paper surface in the annulus family'. Notice that, in this sense, the chimney is itself an annulus,

$$(P \dotplus Q) +_e (R \dotplus S).$$

An engineer would call the square base in Fig. 4.5(*a*) a 'flange'. Thus we have seen that an annulus with a flange is still an annulus.

The basic feature in our description of the punctured torus on p. 21 was that it could be expressed as

$$A +_b N, \tag{2}$$

where A is an annulus (i.e. A is in Family (annulus)) and N is a lamina which is joined as a bridge to A. (The subscript 'b' added to the plus sign is meant to indicate that N is added as a bridge.) Thus, for example, Fig. 4.6

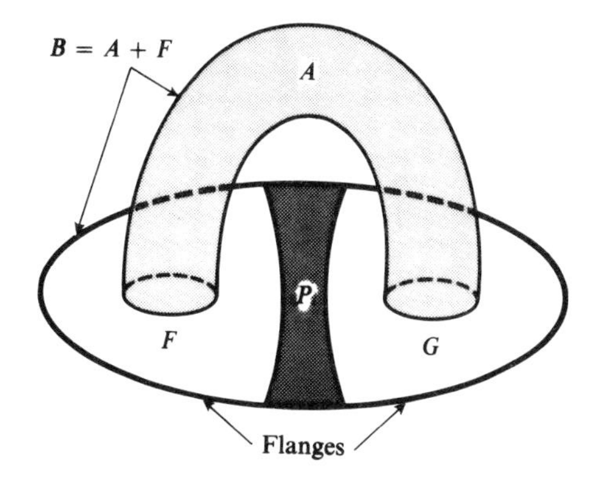

Fig. 4.6

also represents a paper surface S which belongs to the Family (punctured torus), for if we mark in the shaded panel P, we see that the unshaded portion is an annulus A with flanges F, G added at each end. Therefore $A+F$ is an annulus B, being an annulus-plus-flange; and therefore A plus the two flanges is $B+G$, an annulus-plus-flange. This shows that $A+F+G$ is again an annulus, say C. Therefore Fig. 4.6 represents a paper surface which we can express as

$$C +_{\mathrm{b}} P$$

because the panel P is added as a bridge. By comparison with (2) therefore, S is a punctured torus. (We can also think of the surface in Fig. 4.6 as a 'handle' added to P.)

Exercise 4.2

Show that the two surfaces sketched in Fig. 4.7 are punctured toruses.

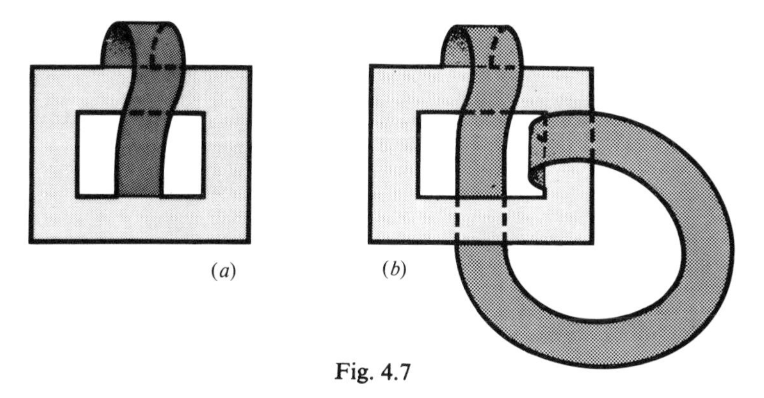

Fig. 4.7

In the same kind of way, we recognize that a paper surface S is a Moebius band, if we can express it in the form

$$S = L +_{\mathrm{te}} M, \tag{3}$$

where L and M are laminas and '$+_{te}$' means that M is added to L as a twisted ear.

And if, instead, we use a twisted bridge L to construct an S of the form

$$A +_{tb} L, \tag{4}$$

where A is an annulus and L is a lamina, then we recognized on p. 21 that S is in the family

Family (punctured Klein bottle).

The expressions (1)–(4) can be regarded as instructions for making the four kinds of surface. If we asked somebody to make a punctured Klein bottle, for example, then he might come back with something of plastic, beautifully curved and shining. However much pleasure we got from the curves and the shine, the only way that we have of deciding whether or not he has made us a punctured Klein bottle, is to see whether formula (4) is satisfied. This might take some thought, because the man might have made his surface look like Fig. 2.4 instead of a lamina with two twisted ears, as in Fig. 2.7. Thus, he would have given us the form (4) rather than the form

$$M \dotplus N, \tag{5}$$

formed by taping together two Moebius bands along an arc of the boundary. The argument on p. 21 allows us to combine (4) and (5) in an equation

$$A +_{tb} L = M \dotplus N, \tag{6}$$

and either form can then be used for recognizing a punctured Klein bottle.

The expressions (1)–(4) allow us to draw a plan Ⓣ for each of the four surfaces, as in Fig. 4.8.

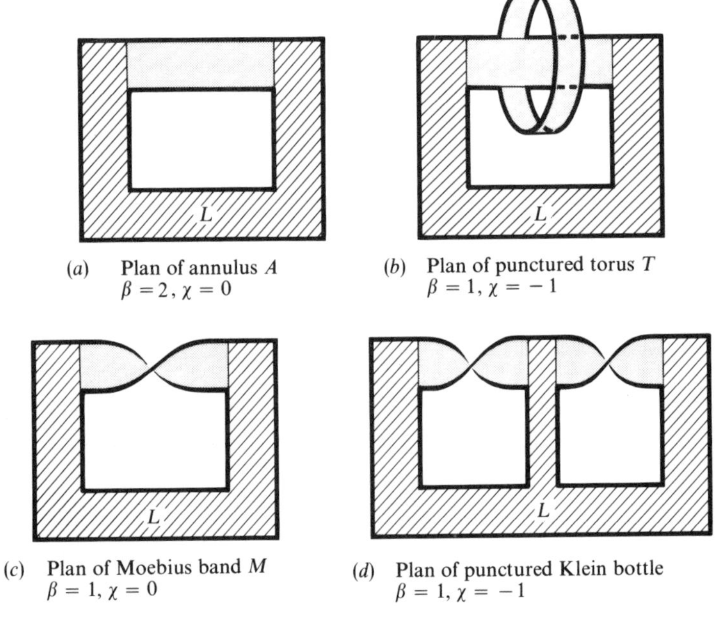

(*a*) Plan of annulus A
$\beta = 2, \chi = 0$

(*b*) Plan of punctured torus T
$\beta = 1, \chi = -1$

(*c*) Plan of Moebius band M
$\beta = 1, \chi = 0$

(*d*) Plan of punctured Klein bottle
$\beta = 1, \chi = -1$

Fig. 4.8

4.3 THE FAMILIES OF PLANAR REGIONS, AND OF SPHERES WITH HANDLES

From the same point of view, all planar surfaces with (say) four holes, have the plan of Fig. 4.9(*a*), because according to our definition on p. 14, such

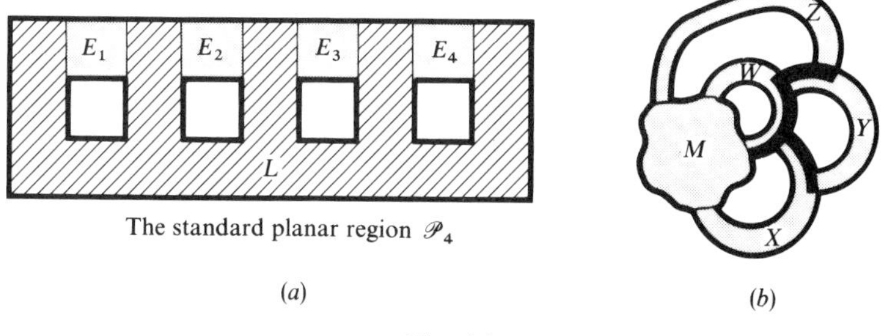

(*a*) (*b*)

Fig. 4.9

a planar surface is made by adding four ears to a panel, so it has an expression of the form – dropping brackets:

$$M +_e A +_e B +_e C +_e D. \tag{7}$$

In Fig. 4.9(*b*) we show how to take the shaded areas away from the ears and add them to the panel M, obtaining a lamina N to which the remains W', X', Y', Z' of the ears are added; each of these new ears is joined to N, just as the ones in the plan are joined to L, and no new ear is now joined to another along an edge.

Analogously, for a planar region with n holes; its plan[Ⓣa] $\mathscr{P}_n$ is like that of Fig. 4.9(*a*) except that L is replaced by a lamina with n (rather than 4) 'bays', and an ear is stretched across each bay, as E_1 is stretched in the picture. Thus $\mathscr{P}_1$ is also the plan for the annulus family.

Now, in our description of Fig. 4.9, we were quietly making two agreements, that we should now be open about, when we removed the shaded areas in Fig. 4.9(*b*) from the ears (we then added them to M, and used Agreement 2). After their removal, the remaining pieces of ear were still agreed to be ears. Thus, we record:

Agreement 4

Let K be a lamina, formed from some of the panels of a lamina D. Suppose K is removed from D to leave a residue N, so that $D = K \dotplus N$. Then N is a lamina.

There was also a second agreement, in that we were not fussy about just how the ears in Fig. 4.9 were stuck on. That agreement can in fact be *deduced* from the others (see Exercise 4.4B, No. 4, below), but for simplicity we shall assume[Ⓣb] it here without proof.

Agreement 5

Let S be a paper surface, and let P be a lamina which is taped to S along arcs γ, δ of boundary curves J and K respectively. Let Q be another lamina, taped*

* γ, δ, λ, μ are the Greek letters gamma, delta, lambda and mu respectively.

to S along arcs λ, μ of J and K respectively, and such that Q is twisted if P is twisted, but otherwise not. Then

$$\text{Family}\,(S+P) = \text{Family}\,(S+Q).$$

N.B. If $J = K$, then P and Q are ears; if $J \neq K$, they are bridges.

As an example, we have Fig. 4.10(*a*) in which the symbol '~' is used to

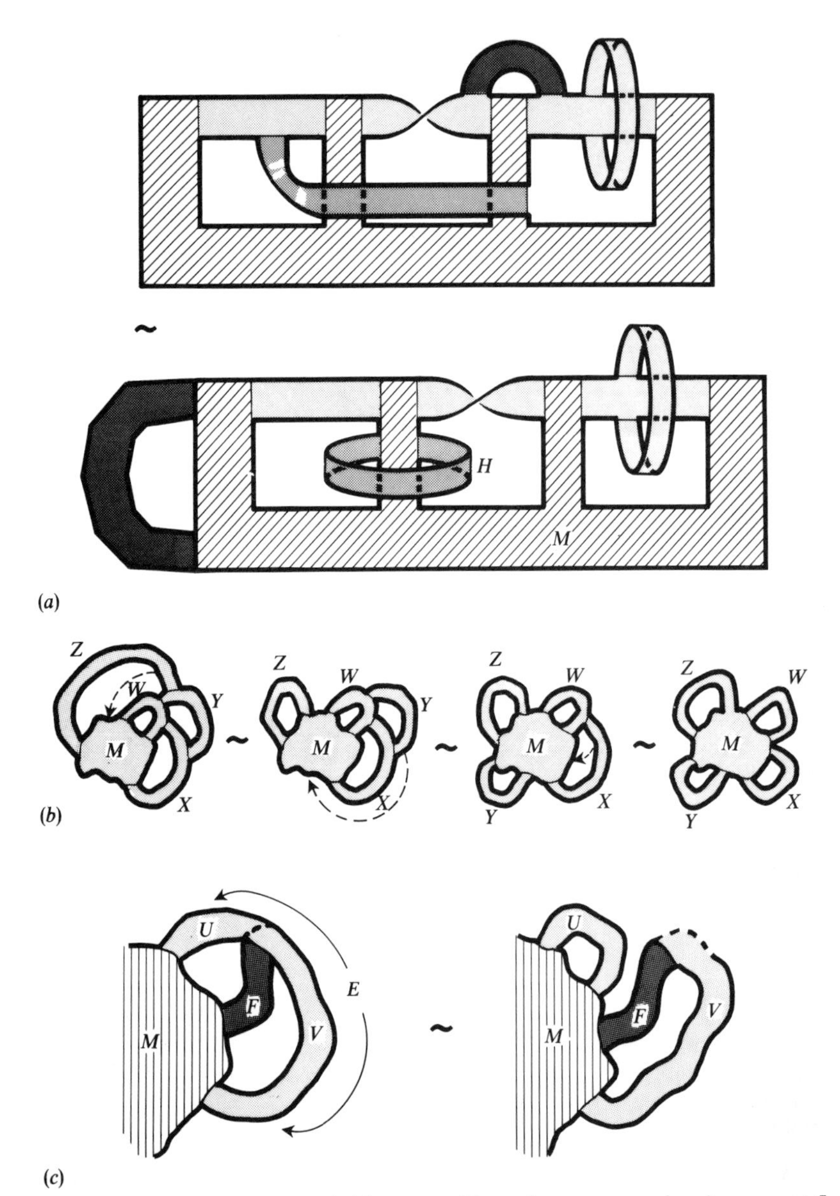

Fig. 4.10. (*a*) Sliding ears and bridges round boundary curves using Agreement 5. (*b*) Sliding ears to put S in the form of plan $\mathscr{P}_4$. (*c*) Ear E repanelled as $E = U \dot{+} V$.

mean 'is in the same family as'. We could deal similarly with the paper surface of Fig. 4.9(b) by using Agreement 5 to rearrange it in the stages displayed by Fig. 4.10(b); the last stage is of the form required by expression (7) above.

Now, an ear might be added across a hole in $\mathscr{P}_n$. In this case, we use Agreement 4 to 'repanel' the surface slightly, as indicated in Fig. 4.10(c), and then we use Agreement 5 to slide the end of the panel U (which is an ear for the surface $M + (F + V)$) on to the boundary of M. Thus

$$(M +_e E) +_e F = (M +_e U) +_e W,$$

where $W = F + V$. The second form is now in the order required by the prescription (7) on p. 39. The sliding process forms the justification required by the Remark on p. 23; it also *proves* Agreement 5 when S is a Moebius band.

Remark. Note that the dark handle in the upper surface of Fig. 4.10(a) could be interwoven through the other holes to make under- and over-passes, as in Fig. 1.7(d). Nevertheless, Agreement 5 says that we can replace it by a simple handle H, as in the second sketch. Similarly, if it had been twisted: if it had, say, 21 left-hand twists, we could replace it with a handle having just one twist, and that right-handed.

Exercise 4.3A

1. Six rectangular holes are cut in a rectangular shoe-box with lid, to make a 'house' with 2 doors and 4 windows. Show that the remaining cardboard has as plan the planar region $\mathscr{P}_5$. (N.B. The whole box itself is in the cocoon family; with one hole it becomes a lamina, and with two it becomes an annulus with plan $\mathscr{P}_1$.)

2. If now one door is blocked up, show that the plan must be changed to $\mathscr{P}_4$.

3. Show that Family ($\mathscr{P}_n$ + lid) = Family ($\mathscr{P}_{n-1}$).

4. Calculate the number β of boundary curves of $\mathscr{P}_n$ and the Euler number $\chi(\mathscr{P}_n)$.

5. Show that if E in Fig. 4.10(c) is a lamina (sum of several panels), then we can replace it by a single panel, before splitting it into U and V. Show that neither of these changes affects the numbers β and χ.

6. Let A, B be planar regions with 3 holes and 5 holes respectively, and suppose that the outer boundary of A is taped to an inner boundary of B. Find a prescription for the resulting surface, and show that it has plan $\mathscr{P}_7$. (Agreement 5 is not needed here.)

A similar kind of plan to that for $\mathscr{P}_n$ can be given for the punctured sphere with four handles, described on p. 26. There, 4 punctured toruses were each joined by an edge to a panel. In Fig. 4.11, a bridge has been added

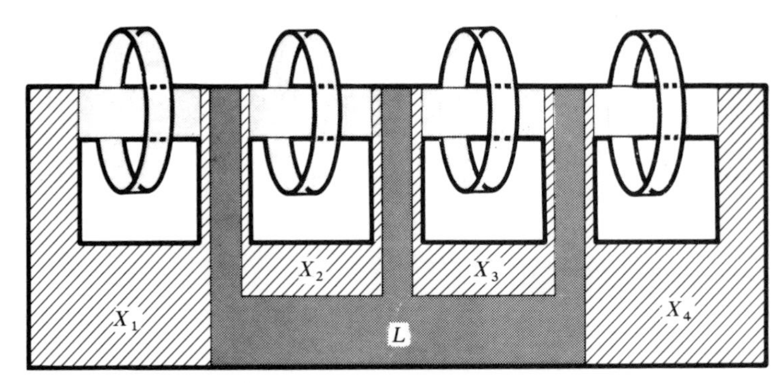

Fig. 4.11. The standard plan $\mathscr{T}_4$.

across each ear of the standard planar region $\mathscr{P}_4$; such a bridge-plus-ear will be called a **handle**. Then we see that the darker area is a lamina to which four punctured toruses X_1, X_2, X_3, X_4 have been attached:

$$\mathscr{T}_4 = (L \dotplus X_1) \dotplus X_2 \dotplus X_3 \dotplus X_4.$$

Any surface with this plan is in Family ($\mathscr{T}_4$) and is called, for brevity, a punctured sphere with 4 handles. It should now be clear what we mean by $\mathscr{T}_2$ or $\mathscr{T}_n$. Note that $\mathscr{T}_1$ is illustrated in Fig. 3.5(a), and $\mathscr{T}_0$ is just a panel.

We expressed $\mathscr{T}_4$, above, as a sum of punctured toruses. It is more convenient, for some purposes, to regard $\mathscr{T}_4$ as a lamina to whose boundary 4 handles are attached (thus ignoring the shading in Fig. 4.11). Thus we may take as the prescription for $\mathscr{T}_n$, the sum

$$M + H_1 + H_2 + \cdots + H_n,$$

where M is a lamina and the H's are handles. We emphasize that M is *any* lamina, not necessarily shaped like the one in Fig. 4.11; and each handle H is composed of two laminas, not necessarily 4-sided (as drawn in the standard version, Fig. 4.11).

Exercise 4.3B

1. Calculate $\beta(\mathscr{T}_4)$, $\chi(\mathscr{T}_4)$, $\beta(\mathscr{T}_n)$, $\chi(\mathscr{T}_n)$.

2. Two handles are added to a base by curved ribs as shown in Fig. 4.12. Show that the resulting paper surface is a punctured sphere with 2 handles, by showing that it has $\mathscr{T}_2$ as its plan.

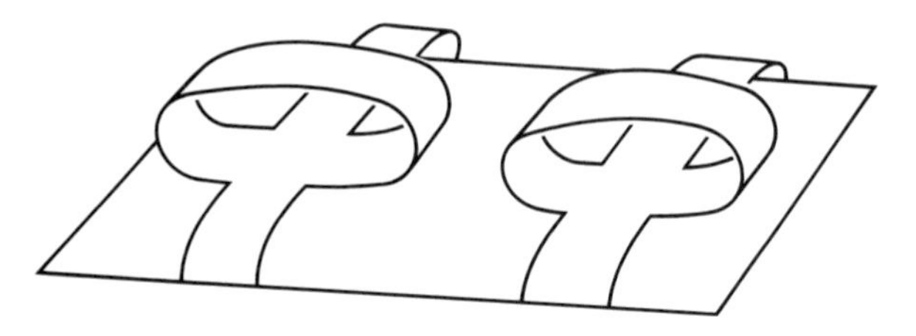

Fig. 4.12

3. Fig. 4.13 shows the surface obtained by punching four square holes in a rectangular block of wood. (It is a closed surface.) By removing the two ends and top, we obtain a surface S with boundary. Show that S is in Family ($\mathscr{T}_4$). [Hint: Use the idea of the previous exercise, observing that the square tunnels are annuli with flanges at each end, when the vertical 'tongues' shaded are regarded as part of the 'floor'.]

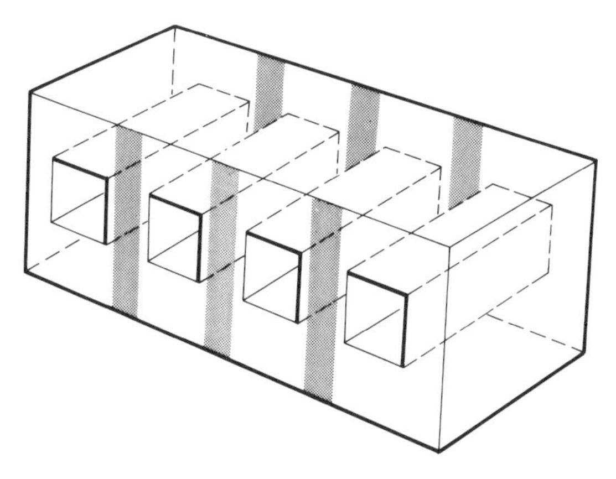

Fig. 4.13

On p. 27 we described a closed surface called the 'sphere with g handles'. It is now reasonable to take as its plan '$\mathscr{T}_g$ plus lid', i.e.

$$\text{Family (sphere with } g \text{ handles)} = \text{Family } (\mathscr{T}_g \mathring{+} L),$$

where L is a lamina. Since $\mathscr{T}_0$ is just a panel, we see that when $g = 0$ the above equation is just the one for the cocoon family on p. 34; a sphere with no handles is a cocoon.

Exercise 4.3C

1. Show[T] that the surface of the 4-tunnelled block, in Fig. 4.13, is a sphere with 4 handles.

2. A cartwheel has sixteen spokes. Show that the surface of spokes, rim and hub, is a sphere with 17 handles. (N.B. The hub has a hole through it!)

3. Recall that a sphere with g handles is said to have genus g. What is the genus of the surface of (i) a Moebius band made of thick paper, (ii) an amoeba, (iii) an earthworm, (iv) a solid earthenware bowl, (v) a human skull?

4. Calculate the number χ for the sphere with g handles. (Since the surface is closed, the number β is zero.)

4.4 THE GENERAL PLAN $\mathscr{S}_{p,q,r}$

In order to discuss the most complicated surfaces with boundary, let us now allow for a plan in which holes, bridges and twisted ears all occur together. (We shall see shortly that twisted bridges can be expressed using two twisted ears.) Such a plan is shown in Fig. 4.14, where we fill in the

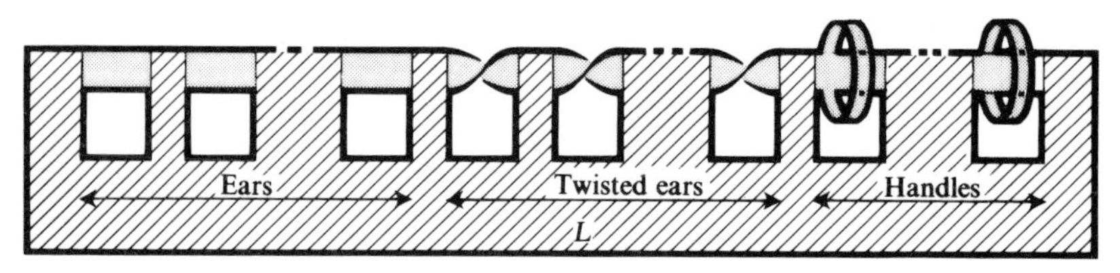

Fig. 4.14. The general plan $\mathscr{S}_{p,q,r}$.

'bays' in the lamina L with a row of (say) p ears, then a row of (say) q twisted ears, and finally a row of (say) r handles. The dots indicate the middle section of each row, since we cannot complete the drawing unless we are given specific numbers for p, q and r. The resulting surface depends on p, q and r, so we build them into its name[T]: $\mathscr{S}_{p,q,r}$ (read '$\mathscr{S}$-pee-ku-ar'); thus, for example, $\mathscr{S}_{10,19,35}$ is our name for the plan that has ten ears, nineteen twisted ears, and thirty-five handles. Also, referring back to the previous plans we have described, we see that $\mathscr{S}_{p,0,0} = \mathscr{P}_p$, $\mathscr{S}_{0,0,r} = \mathscr{T}_r$, $\mathscr{S}_{0,1,0}$ is the Moebius band and $\mathscr{S}_{0,2,0}$ the punctured Klein bottle. In

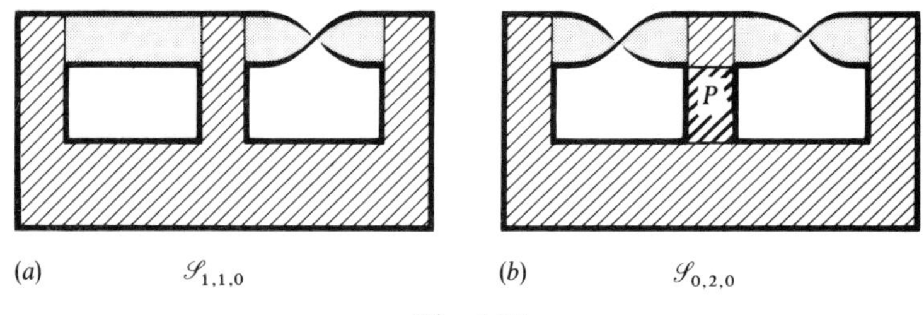

(a) $\mathscr{S}_{1,1,0}$ (b) $\mathscr{S}_{0,2,0}$

Fig. 4.15

Fig. 4.15, we show the difference between $\mathscr{S}_{0,2,0}$ and $\mathscr{S}_{1,1,0}$, the latter being a Moebius band with a hole in it. Clearly, $\mathscr{S}_{0,0,0}$ will mean a rectangular panel, since it certainly has no holes, twists or handles.

Exercise 4.4A

1. Three holes are punched in a punctured Klein bottle, and in a punctured torus. Find plans $\mathscr{S}_{p,q,r}$ for each resulting surface. Now tape one end of a cylinder to one boundary curve of one of the surfaces, and tape its other end to a boundary curve of the other surface. Find a plan $\mathscr{S}_{p,q,r}$ for the new surface.

2. Calculate the number β of boundary curves of $\mathscr{S}_{1,2,3}$ and $\mathscr{S}_{3,2,1}$. Calculate also their Euler numbers χ (see p. 18).

3. Calculate the change in $\beta(\mathscr{S}_{p,q,r})$ and $\chi(\mathscr{S}_{p,q,r})$ when we increase p, q, or r by 1. Hence show that $\beta(\mathscr{S}_{p,q,r}) = 1 + p$, $\chi(\mathscr{S}_{p,q,r}) = 1 - (p + q + 2r)$.

4. Show that if $p > 0$ and the furthest left-hand ear E of $\mathscr{S}_{p,q,r}$ is removed, then the resulting surface T is in Family $(\mathscr{S}_{p-1,q,r})$.

5. If E and T are as above, suppose the 'window' below E is filled in with a lamina P. Show that the resulting surface is in Family $(\mathscr{S}_{p-1,q,r})$.[Hint: $S + P$ can be written $(T \dotplus P) + E$, where $S = \mathscr{S}_{p,q,r}$; then use Agreement 1.]

6. Show that $\mathscr{S}_{p,0,r}$ is orientable, by assigning arrows to its panels.

The paper surface drawn in Fig. 4.14 is what we mean by $\mathscr{S}_{p,q,r}$, and we shall use it as a representative of its Family that we can actually point to, as we might pick a tetrahedron as a familiar representative of the cocoon Family (more familiar than, say, the icosahedron). We are not very interested in the size of $\mathscr{S}_{p,q,r}$, nor in the various lengths of its parts, nor in its colours. Our primary interest is in the formula for putting $\mathscr{S}_{p,q,r}$ together, and this is given by the prescription[T]

$$M + (E_1 + \cdots + E_p) + (F_1 + \cdots + F_q) + (H_1 + \cdots + H_r), \qquad (8)$$

where M is any lamina, the E's are untwisted ears, the F's are twisted ears, and the H's are handles, and where the two ends of each E, F or H are attached *to the boundary of* M. (For example, the handle H in Fig. 4.10 is *not* attached in this way.) We emphasize: M need not have the same number of edges as the lamina L in Fig. 4.14, nor do the E's, F's and H's need to consist of 4-sided panels. Whenever they do not, however, we may replace them by such panels, by our various Agreements. The resulting prescription would not have changed, in the sense that the new panels are still ears or bridges, fitting with the others in just the same relationships as the laminas they replace. And any paper surface S, whose panels can be grouped into laminas $M, E_1, \ldots, E_p, F_1, \ldots, F_q, H_1, \ldots, H_r$ that fit together as in (8) above, is then in Family $(\mathscr{S}_{p,q,r})$. We shall say that (8) is the *prescription* of Family (S), and its plan is $\mathscr{S}_{p,q,r}$. For brevity, we write $U \sim V$ to mean that the paper surfaces U and V are in the same Family; thus, for example, $S \sim \mathscr{S}_{p,q,r}$.

As another example, we have

$$\mathscr{S}_{p,q,r} \sim \mathscr{S}_{p,0,0} \dotplus \mathscr{S}_{0,q,r} (= \mathscr{P}_p \dotplus \mathscr{S}_{0,q,r}),$$

since the sum may be written (using (8)):

$$[(E_1 + \cdots \dotplus E_p) + M] \dotplus N + (F_1 + \cdots + F_q) + (H_1 + \cdots + H_r)$$
$$= Q + (E_1 + \cdots + E_p) + (F_1 + \cdots + F_q) + (H_1 + \cdots + H_r) \sim \mathscr{S}_{p,q,r},$$

where $Q = M \dotplus N$ is a lamina by Agreement 2, because M and N are laminas.

Exercise 4.4B

1. Show that

$$\mathscr{S}_{p,q,r} \sim \mathscr{S}_{p,q,0} \dotplus \mathscr{S}_{0,0,r}$$
$$\sim \mathscr{S}_{p,0,r} \dotplus \mathscr{S}_{0,q,0}.$$

2. Show that if a surface has the prescription (8) above, then its numbers β and χ are the same as if we worked them out using the laminas of the prescription (instead of its panels).

3. Without using Agreement 5, show that if T is an annulus and Q a panel which is added to T as an ear, then Family $(T + Q)$ is that of $\mathscr{S}_{1,1,0}$ if Q is twisted, and that of $\mathscr{S}_{2,0,0}$ if not. If, instead, T is a $\mathscr{P}_2$, show that $T + Q \sim \mathscr{S}_{2,1,0}$ or $\mathscr{S}_{3,0,0}$ according as Q is twisted or untwisted.

***4.** Now deduce Agreement 5 from the others as follows. If the boundary curves J, K of S are distinct, thicken them slightly to make annuli, and join these by an untwisted lamina in S, to make from them a surface X which is a $\mathscr{P}_2$. Let

$S = T + X$, where T is what remains of S after X is removed; thus the entire outer boundary of X is joined to a boundary curve of T. Now let laminas P, Q be added to S along J and K as in the statement of Agreement 5. Then $S + P = T + (X + P)$, and by the previous question, $X + P \sim X + Q$. Hence show that $S + P \sim S + Q$ as required. The case when $J = K$ is handled similarly, but using an annulus for X. [Use Equation (8) on p. 45.]

4.5 THE TRADING THEOREM

We shall now prove a very surprising thing, which can be expressed as a theorem (the reason for whose name will become clear in Section 5.1; see also Exercise 4.5 below).

The Trading Theorem

$$\text{Family } (\mathscr{S}_{0,3,0}) = \text{Family } (\mathscr{S}_{0,1,1}).$$

Before proving this, let us think about what it is telling us.

The surface $\mathscr{S}_{0,3,0}$ is the result of attaching a twisted ear to an $\mathscr{S}_{0,2,0}$, i.e. to a punctured Klein bottle K (see Fig. 4.15(b)). The surface $\mathscr{S}_{0,1,1}$ is the result of attaching a twisted ear to an $\mathscr{S}_{0,0,1}$, i.e. to a punctured torus T. Now the surface $\mathscr{S}_{0,2,0}$ in Fig. 4.15(b) is also the result of adding the darker panel P to the rest, i.e. to a 2-twisted annulus. In the diagram, P looks like a bridge but, by Remark 2 on p. 31, it is twisted. Clearly $\mathscr{S}_{0,0,1}$ is an annulus with bridge shaded in Fig. 3.5(a). Although a 2-twisted annulus belongs to the annulus family (because its plan is $\mathscr{S}_{1,0,0}$), it turns out that we cannot undo the twists, if the panel P is present, without tearing P irreparably. We cannot prove this last statement yet (see Exercise 7.5, No. 4), but it is saying that

$$\begin{cases} \text{Family (2-twisted annulus)} = \text{Family (annulus)}, \\ \text{Family } (P + \text{2-twisted annulus}) \neq \text{Family (bridged annulus)}. \end{cases}$$

And then the statement of the above theorem says:

Family ((P + 2-twisted annulus) $+_t$ ear) = Family (bridged annulus $+_t$ ear).

To prove the theorem, we let[Ⓣa] K' and T' respectively denote the results $\mathscr{S}_{0,3,0}$ and $\mathscr{S}_{0,1,1}$, of adding a twisted ear to K and to T, and then we study Fig. 4.16 (which[Ⓣb] did not arise out of the blue, but only after a lot of experiment with paper models).

In Fig. 4.16(a) we have illustrated the plan of $\mathscr{S}_{0,3,0}$ consisting of a lamina L and three twisted ears across the 'bays' of L. Two squares have been removed from L to leave the (shaded) lamina N, and the squares have been added to the three ears to make the tinted region. This has then been divided into the three laminas P, Q, R. Then Q inherits a twist from the middle ear, so it forms with N the Moebius band $N +_{te} Q$. The two laminas P and R then form *twisted* ears so that

$$\mathscr{S}_{0,3,0} = (N +_{te} Q) +_{te} P +_{te} R, \tag{9}$$

a Moebius band with two twisted ears; each is twisted because neither affects the number of boundary curves, when it is added to the rest.

In Fig. 4.16(b) we have illustrated the plan of $\mathscr{S}_{0,1,1}$, which we have then subdivided into the two shaded laminas and the lighter portion. The latter is clearly a Moebius band M because it incorporates the original

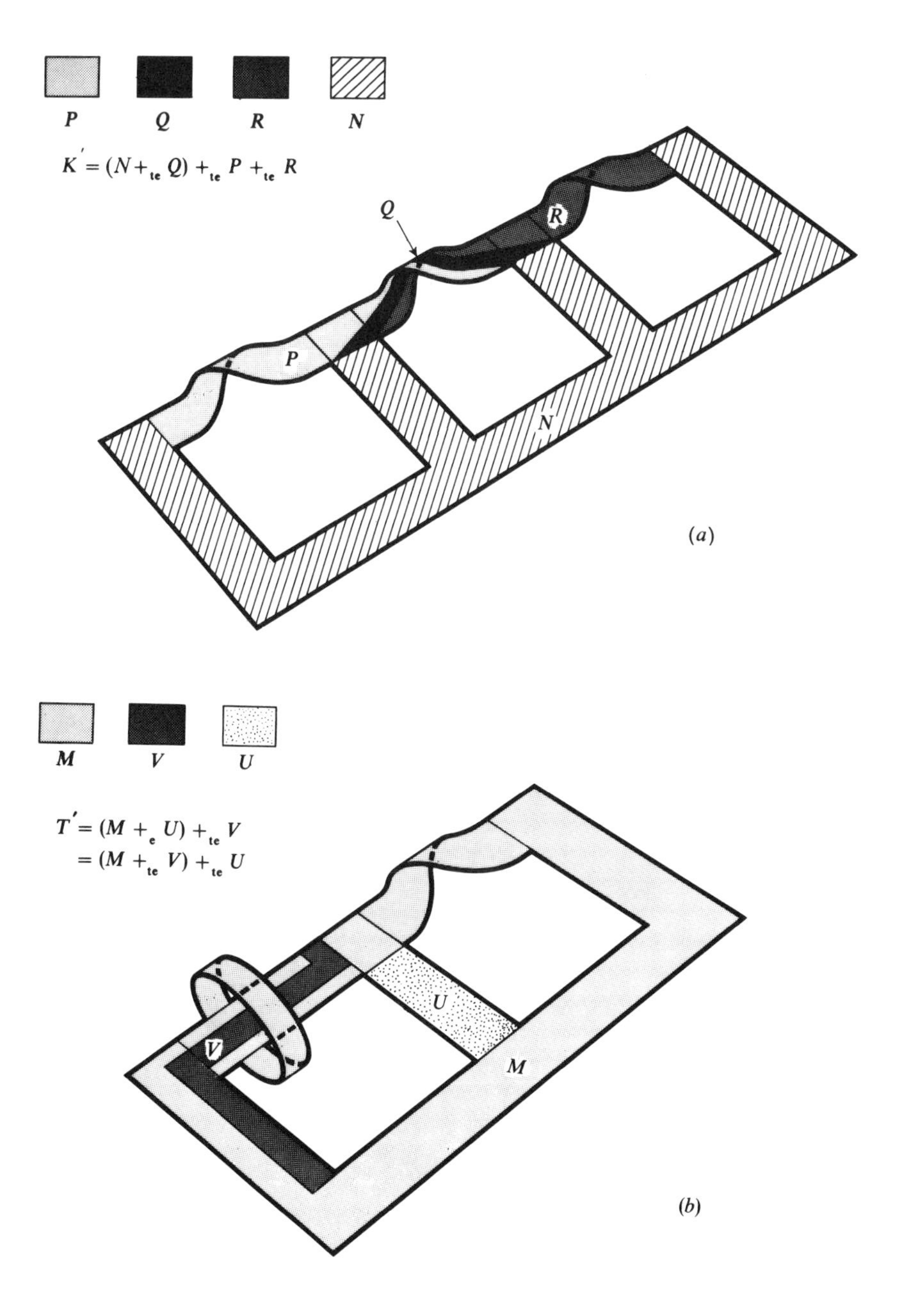

Fig. 4.16. (*a*) $N + Q$ is a Moebius band; P and R are twisted ears. (*b*) M is a Moebius band; U and V are twisted ears.

twisted ear. Now $M + U$ has two boundary curves, so U is an untwisted ear in $M + U$. On the other hand, $M + V$ has *one* boundary curve, so V is twisted in $M + V$; and then U is twisted in $(M + V) + U$. Therefore

$$\mathscr{S}_{0,1,1} = (M +_{te} V) +_{te} U, \tag{10}$$

again a Moebius band with two twisted ears. Therefore (see Remark, p. 23)

$$\text{Family}\ (\mathscr{S}_{0,3,0}) = \text{Family}\ (\mathscr{S}_{0,1,1})$$

and this completes the proof of our theorem.

Exercise 4.5

Show that
$$\begin{aligned}\text{Family}\ (\mathscr{S}_{0,4,0}) &= \text{Family}\ (\mathscr{S}_{0,2,1}),\\ \text{Family}\ (\mathscr{S}_{0,5,0}) &= \text{Family}\ (\mathscr{S}_{0,1,2}),\\ \text{Family}\ (\mathscr{S}_{0,6,0}) &= \text{Family}\ (\mathscr{S}_{0,2,2}),\\ \text{Family}\ (\mathscr{S}_{0,7,0}) &= \text{Family}\ (\mathscr{S}_{0,1,3}), \text{etc.}\end{aligned}$$
(Use the result of the last theorem.)

4.6 ADDING A LAMINA TO $\mathscr{S}_{p,q,r}$

You may have been worried, in the proof of the theorem in the last section, that we started with the plans $\mathscr{S}_{0,2,0}$ and $\mathscr{S}_{0,1,1}$, whereas twisted ears might have been added less neatly to K and T. We shall now deal with such objections, in the course of a much longer argument.

Let us consider what happens if we add a lamina P to a paper surface $\mathscr{S}_{p,q,r}$. The simplest way of adding P is to form $\mathscr{S}_{p,q,r} + P$, but by Agreement 1 this is in Family ($\mathscr{S}_{p,q,r}$). The other ways of adding P are as a lid, an ear, a twisted ear, a bridge or a twisted bridge (since multi-ears can be formed from a succession of simpler additions, as we saw in Section 3.1). We shall now investigate what happens, and it will help the reader if he knows in advance what we shall find. Our findings can be conveniently summarized as a theorem, in which we leave the case of a lid until later. Although we state the theorem and then prove it, this is exactly opposite to the process of investigation, where combinations of guessing and experiment show what can be proved and then what should be formulated.

The Addition Theorem

If, to the plan $\mathscr{S}_{p,q,r}$, we add a new ear, twisted ear, bridge or twisted bridge in any manner (joining along any boundaries), the resulting paper surface has plan

$$\mathscr{S}_{p+1,q,r},\quad \mathscr{S}_{p,q+1,r},\quad \mathscr{S}_{p-1,q,r+1},\quad \mathscr{S}_{p-1,q+2,r}$$

corresponding to each of the four cases. (*In the last two, $p \geqslant 1$.*)

Proof. The proof consists of looking at the various cases.Ⓣ

Case 1. Adding an untwisted ear. If P is added to $\mathscr{S}_{p,q,r}$ as an untwisted ear, we saw with Fig. 4.10, that Agreement 5 allows us to add P as in Fig. 4.17(a) – if P is to be attached to the outer boundary of $\mathscr{S}_{p,q,r}$. But if we attach P to an inner boundary (which is possible only when $p \geqslant 1$), then we can do what we did in Fig. 4.10: we split an ear into three panels as in Fig. 4.17(b). With either possibility then, we see that we have a prescription for $\mathscr{S}_{p,q,r} + P$ of the form (8) on p. 45, but with $p+1$ ears. Therefore

$$\text{Family}\ (\mathscr{S}_{p,q,r} +_{\text{e}} \text{panel}) = \text{Family}\ (\mathscr{S}_{p+1,q,r}).$$

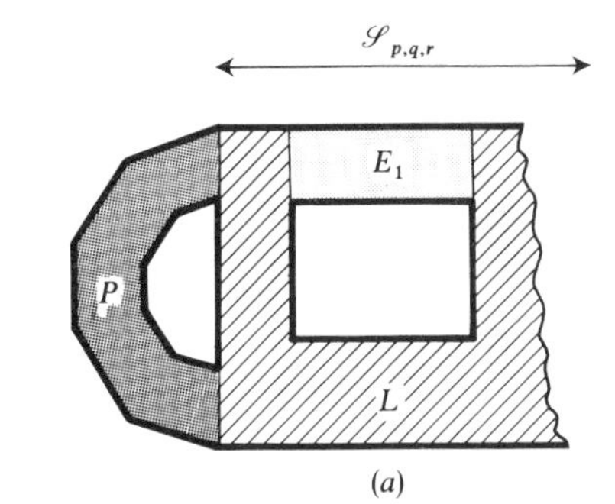

(a)

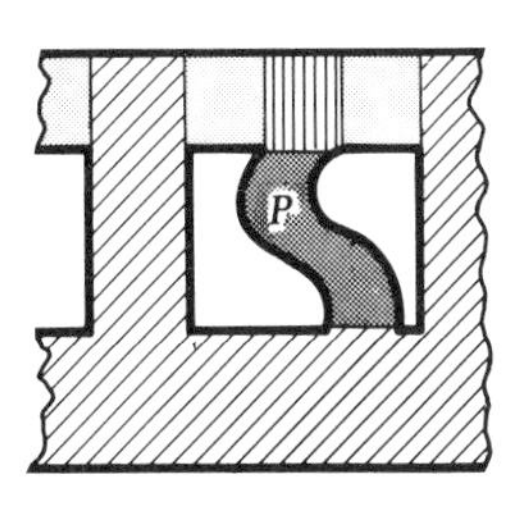

~

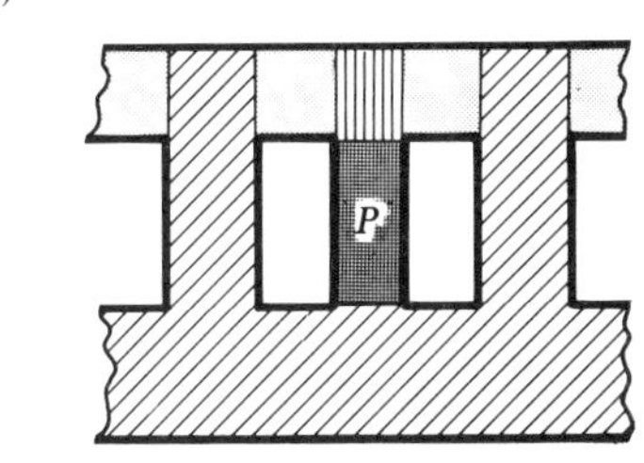

(b)

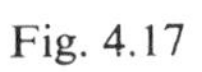

Fig. 4.17

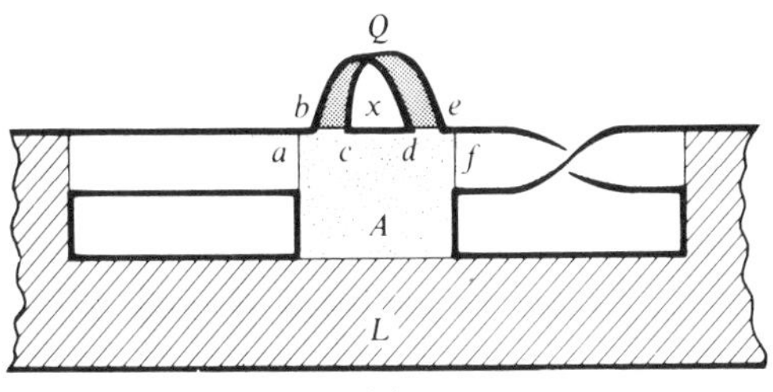

(a)

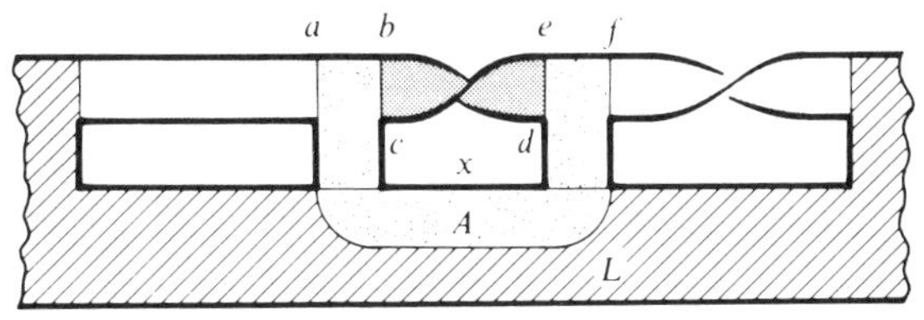

(b)

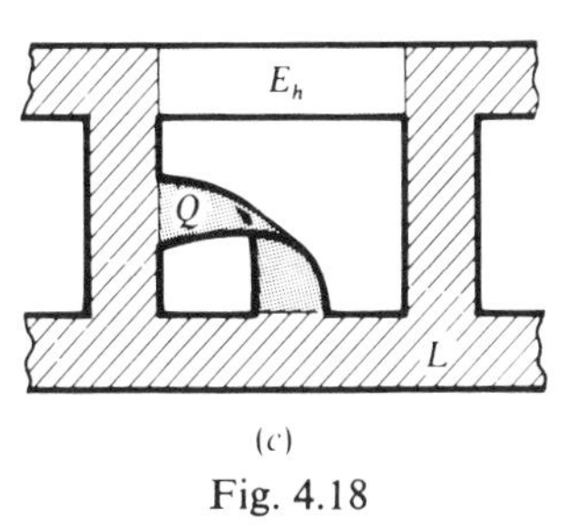

(c)

Fig. 4.18

Case 2. Adding a twisted ear. By Agreement 5, we can choose a very narrow lamina Q, and add it to the 'stump' A of the lamina L, that separates the last of the untwisted ears of $\mathscr{S}_{p,q,r}$ from the first of the twisted ears. (We could then 'dent' A as shown in Fig. 4.18(*b*), to make Q lie neatly across it like the other twisted ears of $\mathscr{S}_{p,q,r}$.) And now Q is the first twisted ear of $\mathscr{S}_{p,q+1,r}$, so we have shown

$$\text{Family}\,(\mathscr{S}_{p,q,r} +_{\text{te}} \text{panel}) = \text{Family}\,(\mathscr{S}_{p,q+1,r}).$$

On the other hand, if we add Q to the *inner* boundary of $\mathscr{S}_{p,q,r}$ then by Agreement 5 we may suppose that it is added as shown in Fig. 4.18(*c*), corresponding to the second possibility in Case 1. If T is the result of removing the ear E_h from $\mathscr{S}_{p,q,r}$ then $\text{Family}\,(T) = \text{Family}\,(\mathscr{S}_{p-1,q,r})$; hence, since Q is added to the outer boundary of T then

$$\text{Family}\,(T +_{\text{te}} Q) = \text{Family}\,(\mathscr{S}_{p-1,q+1,r}),$$

by the first part above. Therefore, since E_h is added to T as an ear, then

$$\begin{aligned}\text{Family}\,(\mathscr{S}_{p,q,r} +_{\text{te}} Q) &= \text{Family}\,((T +_{\text{te}} Q) +_{\text{e}} E_h)\\ &= \text{Family}\,(\mathscr{S}_{p-1,q+1,r} +_{\text{e}} E_h)\\ &= \text{Family}\,(\mathscr{S}_{p,q+1,r})\end{aligned}$$

by Case 1 above.

Thus, however we add the twisted ear, we have shown

$$\text{Family}\,(\mathscr{S}_{p,q,r} +_{\text{te}} \text{panel}) = \text{Family}\,(\mathscr{S}_{p,q+1,r}).$$

Case 3. Adding a bridge. If we add a lamina P to $\mathscr{S}_{p,q,r}$, as a bridge, then we must have $p > 0$ in order to have at least two boundary curves. Then there are two possibilities:

(i) one end of P is on the outer boundary of $\mathscr{S}_{p,q,r}$, and the other end is on an inner boundary curve;

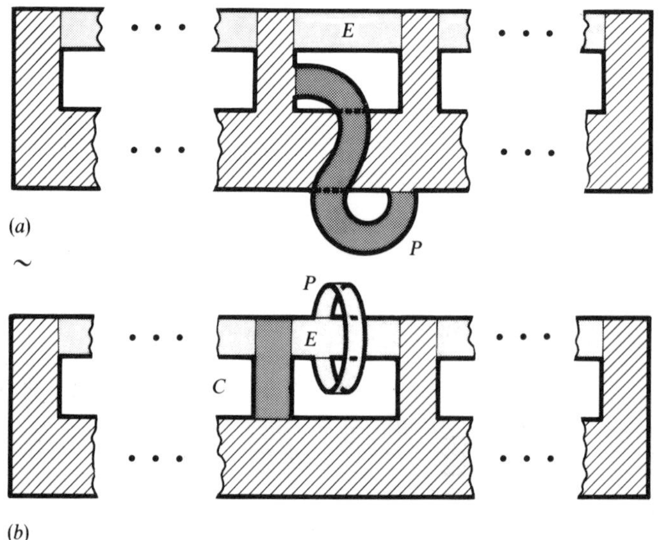

Fig. 4.19

(ii) the ends of P are on (different) inner boundary curves.

Now on $\mathscr{S}_{p,q,r}$, inner boundary curves arise only from the untwisted ears, so by Agreement 5 we may suppose that with possibility (i), the two ends of P are on opposite sides of the same ear (see Fig. 4.19(*b*)). We can then slide any ears, that may lie to the right of $E+P$, onto the 'stump' C in Fig. 4.19(*b*), just as in Cases 1 and 2 above. In this way, $E+P$ becomes the leader of the row of handles already on $\mathscr{S}_{p,q,r}$; and since E has been lost from the set of p ears we have the plan $\mathscr{S}_{p-1,q,r+1}$, for $\mathscr{S}_{p,q,r}+P$. In possibility (ii) above, the ends can lie as shown in Fig. 4.20(*a*) with a certain number of ears between the two 'stumps' A and B to which P is attached. The ears E, F must be untwisted, since inner boundary curves arise only from untwisted ears. But then we can slide each of the untwisted ears, except E and F, to the left, as in Case 1 above, onto the 'stump' C;

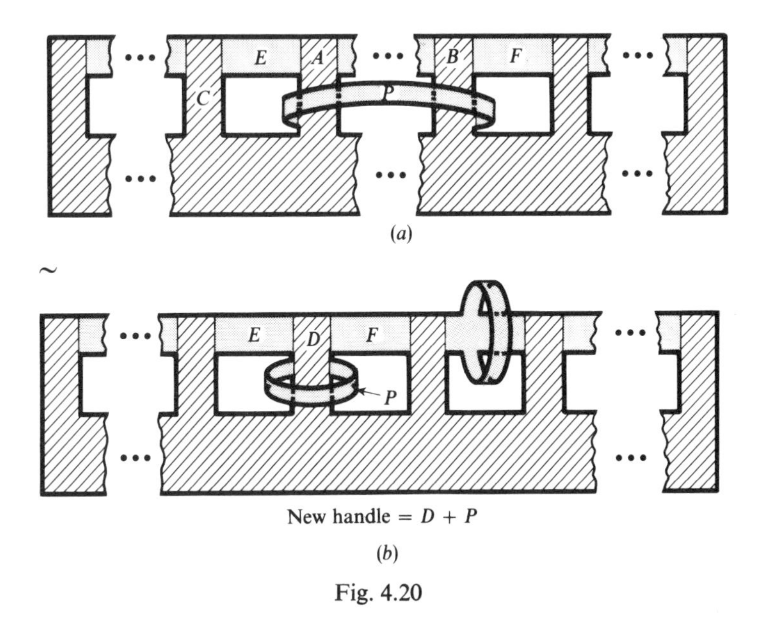

Fig. 4.20

and then P forms a handle H attached to one stump, into which A and B are incorporated. We now slide all the twisted ears to the left, past the handle until the resulting surface is as in Fig. 4.20(*b*), with only the handles of $\mathscr{S}_{p,q,r}$ to its right. It remains to 'interchange' F and the handle H; and this is shown[T] in Fig. 4.21. (Try it yourself first, to see whether you can hit on the trick.) The black panel Q forms with E an untwisted ear A. Also the shaded panel R forms, with the white part of F, an untwisted ear B. Now $\mathscr{S}_{p,q,r}$ minus E, F, P, Q and R, forms a paper surface T with $p-2$ ears, q twisted ears and r handles. If we tape back B, and the white portion P' of P, we have a new surface M consisting of T plus a new handle formed by the ear B bridged by P'. Then

$$\text{Family}\ (M) = \text{Family}\ (\mathscr{S}_{p-2,q,r+1}).$$

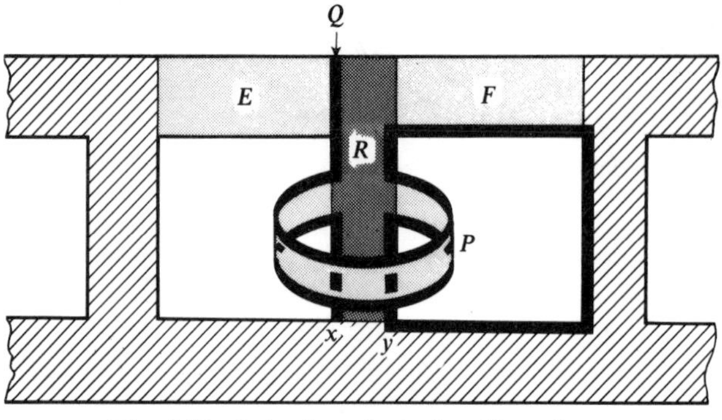

Fig. 4.21. Interchanging a handle and an ear.

Finally, we add back[Ⓣ] the untwisted ear A to get

$$\begin{aligned}\text{Family}\,(\mathscr{S}_{p,q,r} +_{\mathrm{b}} P) &= \text{Family}\,(M +_{\mathrm{e}} A)\\ &= \text{Family}\,(\mathscr{S}_{p-2,q,r+1} + \text{ear})\\ &= \text{Family}\,(\mathscr{S}_{p-1,q,r+1})\end{aligned}$$

as in Case 1.

Case 4. Adding a twisted bridge. This is similar to Case 3, and we must consider situations like those in Fig. 4.19(*b*) and Fig. 4.21, but where P is twisted. In the first case, we see that $\mathscr{S}_{p,q,r} + P = T \dotplus Q$, where Q is $E + P + R$, R is the shaded region in Fig. 4.22 and T is $\mathscr{S}_{p,q,r} - (E + R)$.

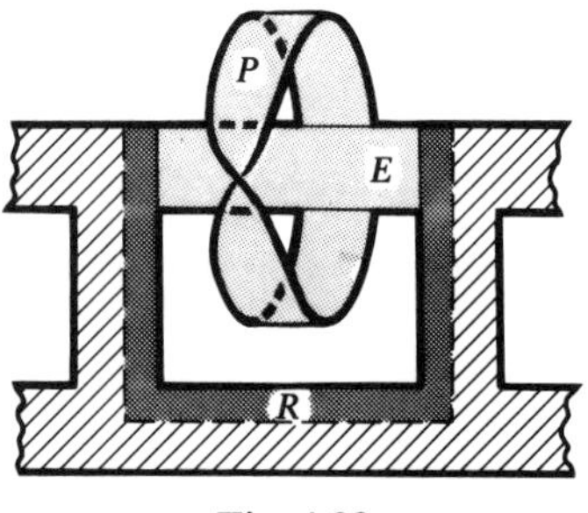

Fig. 4.22

But then Q is a punctured Klein bottle as in Fig. 2.7, with plan $\mathscr{S}_{0,2,0}$. Thus we can replace Q by $\mathscr{S}_{0,2,0}$ to obtain two twisted ears in place of the single ear E, whence

$$\text{Family}\,(\mathscr{S}_{p,q,r} + P) = \text{Family}\,(\mathscr{S}_{p-1,q+2,r}). \tag{11}$$

In the second case, we have a situation as in Fig. 4.21, but in which P is twisted. The same kind of splitting allows us to obtain, as there, an ear modified from E, plus a twisted bridge as in Fig. 4.22. Thus we lose the ear F in Fig. 4.21, and gain two twisted ears as before. Again therefore we have the equation (11) above (but see the Remark on p. 56).

These four kinds of addition therefore have the effect we predicted in the statement of the Addition Theorem, so its proof is complete.

Finally let us consider what happens when we add a lamina to $\mathscr{S}_{p,q,r}$ to form a lid.

Case 5. Adding a lid. Let L be the lamina that forms the lid. There are two possibilities. The first is to join L along one of the inner boundaries of $\mathscr{S}_{p,q,r}$ (which requires $p > 0$). Clearly, the resulting surface is in Family $(\mathscr{S}_{p-1,q,r})$ (for a reason, see Exercise 4.4A, No. 5).

The second way is to join L along the *outer* boundary J of $\mathscr{S}_{p,q,r}$. Again we must consider two possibilities, $p > 0$ and $p = 0$.

First Possibility. $p > 0$. Then

$$\text{Family }(\mathscr{S}_{p,q,r} + \text{lid}) = \text{Family }(\mathscr{S}_{p-1,q,r}).$$

To prove this, note that $\mathscr{S}_{p,q,r}$ possesses an ear because $p > 0$; let D denote its first ear, and let R denote the remainder of $\mathscr{S}_{p,q,r}$. Then $\mathscr{S}_{p,q,r} + L = R + (D + L)$ and D is joined to L along one of two opposite edges. In the taping process we can tape first along this edge to form a panel $M = D \dotplus L$. Then M is taped to R along an arc-succession (see Fig. 4.23), so $R + (D \dotplus L)$ can be written $R \dotplus M$. Therefore

$$\begin{aligned}
\text{Family }(\mathscr{S}_{p,q,r} + L) &= \text{Family }(R + (D + L)) \\
&= \text{Family }(R \dotplus M) \\
&= \text{Family }(R) \qquad\qquad \text{(by Agreement 1)} \\
&= \text{Family }(\mathscr{S}_{p-1,q,r}) \qquad \text{(see Exercise 4.4A, No. 4)},
\end{aligned}$$

so the above claim is justified.

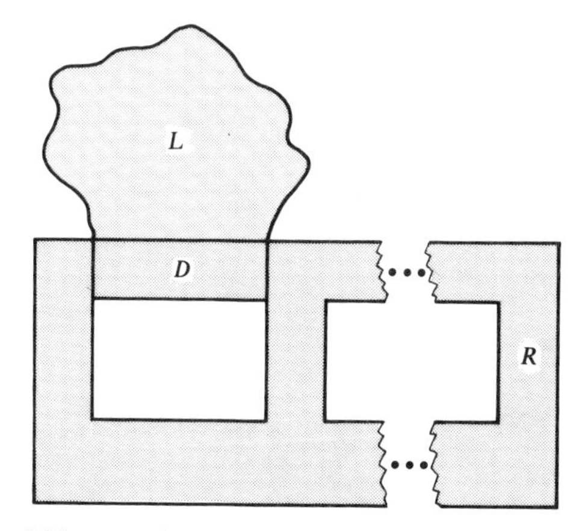

Fig. 4.23. Lid L taped to D along one edge first. $D \dotplus L =$ lamina M.

Second Possibility. If $p = 0$, then J is the only boundary of $\mathscr{S}_{0,q,r}$; and when the lid is added we obtain a *closed* surface. As with the cocoons, we make the following Agreement:

Agreement 6

If Family $(S) =$ Family $(\mathscr{S}_{0,q,r})$, *then*

$$\text{Family }(S + \text{lid}) = \text{Family }(\mathscr{S}_{0,q,r} + \text{lid}).$$

We shall denoteⓉ[a] Family ($\mathscr{S}_{0,q,r}$ + lid) by $\mathscr{C}_{q,r}$. Any surface in $\mathscr{C}_{q,r}$ is called a 'sphere with r handles and q cross-caps'. The name 'cross-cap' is used because we can think of $\mathscr{S}_{0,q,r}$ + lid, as follows. Add a lid to the outer boundary of $\mathscr{S}_{q,0,r}$, to obtain a surface with plan $\mathscr{S}_{q-1,0,r}$ if $q>0$: see Fig. 4.23. For all $q \geqslant 0$, however, the result is a sphere with r handles and q holes. Along the entire boundary of each hole, we now tape a Moebius band, to fill in the hole; when all the holes are filled in like this, the resulting surface T is in Family ($\mathscr{S}_{0,q,r}$ + lid). The standard pictures, of a hole being filled in by a Moebius band, have to be drawn with a 'crossed-over' appearance (see, for example, Hilbert and Cohn-Vossen [9], p. 309), and they lead to the name cross-cap.

Exercise 4.6

1. Give reasons to justify the statement above, that T is in Family ($\mathscr{S}_{0,q,r}$ + lid).

2. Draw a sketchⓉ[b] of a Moebius band that fills a hole. Need your sketch look like the cross-cap of the standard pictures mentioned in the text?

3. Show that $\mathscr{C}_{0,r}$ = Family (closed sphere with r handles), that $\mathscr{C}_{1,0}$ = Family (real projective plane), and that $\mathscr{C}_{2,0}$ = Family (Klein bottle).

4.7 RULES FOR RECOGNIZING SURFACE FAMILIES

Looking back at the work of this chapter, we see that we have been considering certain surfaces and deciding to which families they belong: we have been *classifying* them, just as naturalists recognize certain features of plants or animals in order to classify them as (say) ferns or sharks. In classifying any of these surfaces, we have examined the way in which they were assembled, and then we have used the Agreements 1–6 *and these only* in order to make our classifications. Eventually we want to be in a position to be able to make the following claim about recognizing surfaces:

Recognition Claim

If $S = T + L$, where L is a lamina, then if we can recognize T, and if we know the manner in which L was added to T, then we can recognize S.

So far, we have established the case when T is $\mathscr{S}_{p,q,r}$.

More precisely, Agreement 2 starts us off by saying

$$\text{Family (lamina)} = \text{Family (lamina} \dotplus \text{panel)} = \text{Family (panel)},$$

with the reverse rule (Agreement 4) for subtracting a panel from a lamina. And then Agreements 1, 5 and 6 say that if T is any paper surface and L, M are laminas, then

$$\text{Family } (T +_x L) = \text{Family } (T +_x M), \tag{12}$$

where '$+_x$' stands for the various modes of addition – ($\dotplus$, $\mathring{+}$, $+_e$, etc.). Agreement 3 was a special case of Agreement 6.

Exercise 4.7A

1. Use the Agreements to show that $T +_x L$ and $T +_x M$, in (12) above, have the same number β of boundary curves, and the same Euler numbers.

2. An annulus A is joined along one of its boundary curves, to one boundary curve of a paper surface S. Show that Family $(S + A)$ = Family (S). [Hint: Express A as $L +_e M$, the sum of two laminas, and note that $S + A = (S \dotplus L) \dotplus M$. Use Agreement 1.]

In Section 4.6, we used (12) above, several times, to establish the Recognition Claim for the case when T there is a plan $\mathscr{S}_{p,q,r}$. Let us now extend the Claim a bit further, to the case when T is not $\mathscr{S}_{p,q,r}$ itself, but where we happen to know that T has a plan $\mathscr{S}_{p,q,r}$ (in brief: $T \sim \mathscr{S}_{p,q,r}$).

Thus, according to the discussion on p. 45, we have been able to group the panels of T into laminas, in such a way that T has a prescription

$$T = M + (E_1 + \cdots + E_p) + (F_1 + \cdots + F_q) + (H_1 + \cdots + H_r) \qquad (13)$$

of the form (8) on p. 45; and this prescription corresponds to one for $\mathscr{S}_{p,q,r}$:

$$\mathscr{S}_{p,q,r} = L + (X_1 + \cdots + X_p) + (Y_1 + \cdots + Y_q) + (Z_1 + \cdots + Z_r).$$

Here, M, the E's and F's are laminas composed of panels of T, and the H's are handles, each composed of two laminas (these two being composed of panels of T). On the other hand, L, the X's and the Y's are *panels* of $\mathscr{S}_{p,q,r}$, and each handle Z is composed of two *panels* of $\mathscr{S}_{p,q,r}$.

We now want to recognize $S = T +_x K$, where K is a lamina and '$+_x$' indicates how K is added to T (whether as ear, bridge or lid). Dealing with the case of a lid first, we must look at two possibilities:

(*a*) $p = 0$. Thus $S = T + K$ is a closed surface, and by Agreement 6, $T + K \sim \mathscr{S}_{0,q,r} + \text{lid} = \mathscr{C}_{q,r}$, and we are finished.

(*b*) $p > 0$. Referring back to the First Possibility on p. 53, we follow out the process there described, but instead of the ear D of $\mathscr{S}_{p,q,r}$ that was used there, we repeat the instructions with the ear E of T that corresponds to D. (D was the first ear X_1, so E is the first ear E_1 of D.)

The surface R, that we had there, was recognized as having $\mathscr{S}_{p-1,q,r}$ as its plan; we performed the recognition because we knew the prescription of R (it was that of $\mathscr{S}_{p,q,r}$ minus one ear). In our present case, R is replaced by $T - E_1$; and its prescription is plainly that of $\mathscr{S}_{p-1,q,r}$, as we see by comparing the above prescriptions for T and $\mathscr{S}_{p,q,r}$ and striking out E_1 from the first, and X_1 from the second. Therefore the entire argument, that we applied before, applies to the prescription of T; and we can now conclude with the same type of equalities as before, to get

$$S = T \mathring{+} K \sim \mathscr{S}_{p-1,q,r}.$$

If K were added as an ear or bridge, we work in the same manner, copying each step of the appropriate argument for Cases 1–4 in the last section. (We could begin by using a rectangular panel instead of K, for simplicity, by Agreement 5.) We merely copy Ⓣ with the *laminas* of the prescription of

T, what was previously done with the *panels* and the prescription of $\mathscr{S}_{p,q,r}$.

Therefore we have shown that the Recognition Claim is now established when T there is known to have a plan. In the next chapter we will show that, in fact, T always has a plan. This will then establish the Claim in full generality (see p. 62). So far, however, we can record our conclusions in the form of an extension of (12) on p. 54 as:

Conclusion

If $T \sim \mathscr{S}_{p,q,r}$ *then* $T +_x$ lamina $\sim \mathscr{S}_{p,q,r} +_x$ panel.

Exercise 4.7B

If we look at a solid statue, we note that its surface can be modelled by a closed paper surface S. If, say, an arm touches the torso, then it contributes one or more handles to S. Examine some statues – heads of people, crucifixes, statues in museums or parks – and decide to which family $\mathscr{C}_{0,r}$ the corresponding surface of each belongs. Try to find two very different statues whose surfaces lie in the same family. Have the differences a greater effect upon you than the similarities?

Remark (*twisted bridges again*). On p. 31 we showed how to decide whether or not a bridge P is twisted, when we add it to an *orientable* paper surface S. Clearly, if P is twisted or untwisted, it remains in that state even after we slide its ends round the boundary of S, as in Cases 3 and 4 of the proof of the Addition Theorem.

We now claim that if S is *non*-orientable then *it does not matter whether we call P twisted or not*. For, suppose that $S = \mathscr{S}_{p,q,r}$ with $q \geqslant 1$ and $p > 1$ since P is a bridge. Then the Addition Theorem tells us that if we call P twisted then $S + P \sim \mathscr{S}_{p-1,q+2,r}$ while $S + P \sim \mathscr{S}_{p-1,q,r+1}$ otherwise. But the Trading Theorem tells us that three twisted ears can be exchanged for one handle and a twisted ear. Now, $q + 2 \geqslant 3$, so

$$\mathscr{S}_{p-1,q+2,r} = \mathscr{S}_{p-1,q-1+3,r} \sim \mathscr{S}_{p-1,q,r+1}.$$

Hence, Family $(S + P)$ is the same whether we call P twisted or not. A fortunate state of affairs!

5. Completion of the census of surface families

In this chapter, we shall show that we can make a systematic description of *all* Families of paper surfaces. The description is put together from four generalizations, proved below and which we call the Gallic Theorem, the Fundamental Theorem, and the two Classification Theorems. These tell us

(i) every plan $\mathscr{S}_{p,q,r}$ lies in one of three Families,

(ii) every Family contains just one plan of a special kind,

with similar statements for closed surfaces. Thus, we know each Family in terms of the plan which represents it. It is rather as if we could say of the Scottish Clans that each has a unique member (say the Chief) and if we know that member we know the essential features of all other people in his Clan. We cannot make the 'and if' claim for *people*, of course, since people are more complicated – even in their essential features – than are surfaces.

It will help us if we now start to use symbols rather than words to make our statements: a symbol can represent a package of words and thus we can organize our thoughts by manipulating these packages.

5.1 SOME ALGEBRA

We begin by using algebra to write the conclusions of Steps 1–4 in the proof of the Addition Theorem. These can be written in the following forms, when P is a lamina:

The Addition Equations

$$\begin{aligned}
\text{Family}\,(\mathscr{S}_{p,q,r} +_{\text{e}} P) &= \text{Family}\,(\mathscr{S}_{p+1,q,r}),\\
\text{Family}\,(\mathscr{S}_{p,q,r} +_{\text{te}} P) &= \text{Family}\,(\mathscr{S}_{p,q+1,r}),\\
\text{Family}\,(\mathscr{S}_{p,q,r} +_{\text{b}} P) &= \text{Family}\,(\mathscr{S}_{p-1,q,r+1}),\\
\text{Family}\,(\mathscr{S}_{p,q,r} +_{\text{tb}} P) &= \text{Family}\,(\mathscr{S}_{p-1,q+2,r}).
\end{aligned}$$

In particular, therefore, we get nothing new if we add a twisted bridge; two twisted ears would have achieved the same effect, had we first removed a certain hole.

Also, the Trading Theorem of Section 4.4 allowed us to 'trade' every triple of twisted ears for a twisted ear and a handle. Thus

$$\text{Family}\,(\mathscr{S}_{p,q,r}) = \text{Family}\,(\mathscr{S}_{p,q-2,r+1})$$

if q is at least 3. Hence, from the last of the four equations above, we now have

$$\text{Family}\,(\mathscr{S}_{p,q,r} +_{\text{tb}} P) = \text{Family}\,(\mathscr{S}_{p-1,q,r+1}),$$

provided $q \geqslant 1$ to ensure that $q+2 \geqslant 3$. We now repeat the trading provided $q-2 \geqslant 3$; each exchange reduces the twisted ears by 2, and increases the handles by 1. Eventually(T) we shall end with a surface that has either one, or a pair of twisted ears; of course, if there were only 2, 1 or 0 twisted ears to begin with, then no trading could start. We can summarize these remarks as a theorem, recalling Caesar's description of Gaul.

The Gallic Theorem

Every plan $\mathscr{S}_{p,q,r}$ lies in one of the three families

$$\text{Family}\ (\mathscr{S}_{p,0,r}),\quad \text{Family}\ (\mathscr{S}_{p,1,u}),\quad \text{Family}\ (\mathscr{S}_{p,2,v}),$$

according as q is 0, *odd, or even* (*but non-zero*) *respectively, where u and v depend on q and r.*

Exercise 5.1

★**1.** Show that if q is odd, then $u = r + \frac{1}{2}(q-1)$, while if q is even and non-zero, then $v = r - 1 + \frac{1}{2}q$.

★**2.** If $q > 0$, show that Family $(\mathscr{S}_{p,q,r})$ = Family $(\mathscr{S}_{p,q+2r,0})$.

5.2 THE FUNDAMENTAL THEOREM: AN ANSWER TO THE BIG QUESTION

We shall now answer the Big Question that was raised on p. 31, not by a plain 'Yes' or 'No' but in the following more precise form, which we call the Fundamental Theorem – fundamental because it forms the starting point of any further study of surfaces.

The Fundamental Theorem

Every paper surface with boundary has a plan

This means that if S is a paper surface with boundary, then we can find numbers p, q, r so that $\mathscr{S}_{p,q,r}$ is a plan for Family (S) in the sense we explained on p. 45, and

$$\text{Family}\ (S) = \text{Family}\ (\mathscr{S}_{p,q,r}).$$

Put another way, *every Family has a plan.* (We shall show later how to calculate p, q, and r.)

Proof of the theorem. Recall that a paper surface S is formed by taping together a certain number, say n, of panels according to Rules 1 and 2 (see Section 1.2). Certainly, if $n = 1$, then S is simply a panel with plan $\mathscr{S}_{0,0,0}$, and the theorem holds in this case. Let us therefore suppose $n > 1$, say $n = m + 1$, where m is a whole number $\geqslant 1$.

Then the first m panels form a paper surface T, and T cannot be a closed surface, otherwise we would need to violate Rule 2 in order to tape the nth panel P. Also, since S $(= T + P)$ is also not closed (by the assumptions of the theorem), then P is added to T either as a (possibly twisted) ear, a (possibly twisted) bridge, or along an arc-succession of edges. In the last case, by Agreement 1

$$\text{Family}\ (S) = \text{Family}\ (T \dotplus P) = \text{Family}\ (T),$$

so if T has a plan, say $\mathscr{S}_{u,v,w}$, then $\mathscr{S}_{u,v,w}$ is also a plan for S. With the other four ways of adding P, we could use the Conclusion (p. 56) to find a plan for S because we can find the plan $\mathscr{S}_{u,v,w}$ for T; for example, if P is added as an ear, then that Conclusion tells us that $S = T +_e P$ has a plan $\mathscr{S}_{p,q,r}$ (say); and we can then work out p, q and r from the Addition Equations. Thus

$$\begin{aligned}\text{Family}\,(S) &= \text{Family}\,(T +_e P)\\ &= \text{Family}\,(\mathscr{S}_{u,v,w} +_e P)\,(\text{by the Conclusion (p. 56)})\\ &= \text{Family}\,(\mathscr{S}_{u+1,v,w}) \quad (\text{by the Addition Equations}).\end{aligned}$$

Similar arguments hold when P is a bridge, or twisted. But we can now remove the mth panel Q from T, to get $T = U + Q$, and repeat the argument: if we can find a plan $\mathscr{S}_{a,b,c}$ for U, we can find one for T and hence one for S.

Repeating the process of removing a panel at a time, we finally[T] end with a single panel, which has plan $\mathscr{S}_{0,0,0}$ as we saw above. Starting with this plan, we find a plan for the sum of the first 2 panels, then for the first 3, and so on, until we find a plan for T and hence for S. This completes the proof.

The repetitive process used in the proof has a special formulation, called the *method of mathematical induction*. We cannot explain here why it works, and the interested reader may consult Griffiths and Hilton [7], p. 63. However, the basic feature is that we have a statement about a whole number n, and we label the statement E_n, where here E_n is the statement 'Every paper surface with boundary, formed from n panels, has a plan'. We verified that E_1 held, and showed that if we could prove E_m (i.e. if T has a plan) then we could prove E_{m+1} (that S had a plan). This suffices, *without further discussion*, to establish the statement E_n for all n – by a mathematical theorem about such proofs which is called the *Theorem of Mathematical Induction*. The further discussion we gave above (about repeating and reversing) was merely to lend plausibility for those readers who may not have met such inductive proofs before.

5.3 CALCULATION OF p, q AND r

The Fundamental Theorem now raises the question 'Given a paper surface S with boundary, and plan $\mathscr{S}_{p,q,r}$, can we calculate p, q and r?' We may assume, as in Section 5.1, that q is 0, 1, or 2. Of course, we could work out p, q, and r by going through the inductive process of the above proof, but that procedure is not very efficient. Instead, we proceed as follows:
Our surface S has its Euler number $\chi(S)$ that we described in Section 2.6; thus

$$c = \chi(S) = \text{corners} - \text{edges} + \text{panels},$$

and this can be calculated simply by counting. Also S has the number $\beta(S)$ of boundary curves. If S has plan $\mathscr{S}_{p,q,r}$, then

$$\text{Family}\,(S) = \text{Family}\,(\mathscr{S}_{p,q,r})$$

so $c = \chi(S) = \chi(\mathscr{S}_{p,q,r})$ by Exercise 4.4B, No. 2, so $c = 1 - (p + q + 2r)$ by Exercise 4.4A, No. 3. Note that c is negative except for 3 cases. Firstly, if $p = q = r = 0$, then $c = 1$ and S is a panel. If $p = 1$ and $q = 0 = r$, then

$c = 0$ and S is an annulus. If $q = 1$ and $p = 0 = r$, then $c = 0$ and S is a Moebius band. For brevity, let us denote $\beta(S)$ by b. Then

$$b = \beta(S) = \beta(\mathscr{S}_{p,q,r}),$$

which is $1 + p$ by Exercise 4.4A, No. 3 again. Then b is at least 1. Therefore if we know b – and this can be easily counted, as S is built up, panel by panel – then we know p, because $p = b - 1$. Hence, by the first equation,

$$\begin{aligned} q + 2r &= 1 - p - c = 1 - (b - 1) - c \\ &= 2 - (b + c). \end{aligned}$$

Thus, as 2 and $2r$ are even numbers, then q is even if $b + c$ is even, while q is odd if $b + c$ is odd (in brief, q has the same[Ta] *parity* as $b + c$).

Recall that we decided to assume that q is either 0, 1, or 2. Therefore if $b + c$ is *odd*, then q must be 1, while q is 0 or 2 if $b + c$ is even, in order that the parities agree. In particular, observe that if S is a surface and $b = c = 1$, then S is a lamina. For, by the equations above, $q + 2r = 2 - 2 = 0$; but q and r are not negative, so the only possibility is $q = r = 0$. Since $p = b - 1 = 0$ then S has plan $\mathscr{S}_{0,0,0}$; thus S is a lamina, as claimed above.

Exercise 5.3

1. If S is a paper surface with boundary, and $b = 2$ while $c = -21$, find a plan[Tb] $\mathscr{S}_{p,q,r}$ for S.

2. Show that if a bridge is added to a surface with plan $\mathscr{S}_{p,q,r}$ (twisted or not) then the parity of q does not change.

3. Alternate edges of an octagonal panel are joined to the corresponding edges of another octagonal panel by four rectangular panels, each twisted once. Find a plan for the resulting surface.

5.4 THE UNIQUENESS OF THE PLAN OF A SURFACE

Can a paper surface S have two different plans $\mathscr{S}_{p,q,r}$ and $\mathscr{S}_{u,v,w}$, but with $0 \leqslant q \leqslant 2$ and $0 \leqslant v \leqslant 2$? The answer is[Tc] 'No'. To see this, we look back at the proof of the Fundamental Theorem. There we saw that a plan for the general S could be constructed if we knew a plan for T; for S was a sum of T and a panel P, of the form $S = T +_x P$, where 'x' indicates the method of addition. We saw that the Conclusion (p. 56) together with the Addition Theorem, tell us that if T has plan $\mathscr{S}_{u,v,w}$, then

$$\text{Family }(S) = \text{Family }(T +_x P) = \text{Family }(\mathscr{S}_{u,v,w} +_x P) = \text{Family }(\mathscr{S}_{p,q,r}),$$

where p, q, r are given by the Addition Equations (see Section 5.1) in terms of u, v, w and x. Thus, if the plan for T is unique, so is that for S. But the starting plan, for a panel, *is* unique; it is $\mathscr{S}_{0,0,0}$. Hence by the inductive method of proof, S has a unique plan, $\mathscr{S}_{p,q,r}$ provided $0 \leqslant q \leqslant 2$.

In particular, then, the **orientability number** q is determined for the surface S, subject to being 0, 1 or 2. If $q = 0$, then S is *orientable* (in the sense of Section 3.5) and conversely. For, the Addition Equations show that $q = 0$ when, and only when, no panel of S is added as a twisted ear or bridge; now, the starting panel is orientable, and if T is orientable, then so is T plus the panel P, by Exercise 3.5, No. 6. Hence so is S, by mathematical induction, and if S is orientable, so is T. Therefore the first part shows that if $q = 0$, then S is orientable. It now follows that if q is 1 or 2, then S is *non-orientable* (for then q is not equal to zero) while reversing the argument shows that if S is orientable, then q is zero as asserted.

Referring now to the calculations of Section 5.3, we find the possibilities:

(i) $q = 0, \quad p = b - 1, \quad r = \frac{1}{2}(2 - (b + c))$,
(ii) $q = 1, \quad p = b - 1, \quad r = \frac{1}{2}(1 - (b + c))$,
(iii) $q = 2, \quad p = b - 1, \quad r = -\frac{1}{2}(b + c)$,

thus determining p and r uniquely in terms of b $(= \beta(S))$ and c $(= \chi(S))$. Knowing b and c we also know the parity of $b + c$. If this parity is odd then, as seen above, we know that $q = 1$. But if this parity is even, we cannot decide whether q is zero or 2 without looking also at the *order* in which the panels of S were assembled (to apply the argument about T, used above). We shall return to this point in Chapter 7.

Notice that when $S = \mathscr{S}_{u,v,w}$ and $v = 0$, 1 or 2, the numbers p, q, r thus calculated are u, v, w. This shows that no two plans $\mathscr{S}_{u,v,w}$, $\mathscr{S}_{l,m,n}$ are in the same family (if $v = m$) unless $u = l$ and $w = n$, because they would have to possess the same numbers β and χ. Hence *no paper surface S with boundary can be in two different families*, since it has to lie in the same family as its plan.

One part of our arguments can now be summarized in a theorem, as follows, where we call $\mathscr{S}_{p,q,r}$ a **Gallic** plan if $0 \leqslant q \leqslant 2$.

The Classification Theorem

If S is a paper surface with boundary, then it is in exactly one of the three families, Family $(\mathscr{S}_{p,q,r})$, *of the Gallic Theorem with* $\mathscr{S}_{p,q,r}$ *a Gallic plan.*

5.5 THE FAMILY OF A CLOSED SURFACE

Now let S be a paper surface which is closed. Using the families $\mathscr{C}_{q,r}$ introduced on p. 54, we shall prove a classification theorem for them also.

Theorem
If S is a closed paper surface, then it is in exactly one of the families $\mathscr{C}_{q,r}$, *where* $0 \leqslant q \leqslant 2$ *and* $\mathscr{C}_{q,r}$ *is the family of spheres with r handles and q cross-caps. The number q is zero when, and only when, no panel in the assembly of S is twisted.*

Proof. If S was assembled[T] from m panels, in order, then the last panel P (the mth) must have formed a lid for the rest of S. Therefore the first $(m - 1)$ panels must form a paper surface T with just one boundary curve. Hence, by the Classification Theorem (for surfaces with boundary) that we proved in Section 5.4, there must be a plan of the form $\mathscr{S}_{0,q,r}$ such that

$$\text{Family } (T) = \text{Family } (\mathscr{S}_{0,q,r}),$$

where q is 0, 1 or 2, and r is determined from the formulae of Section 5.4. Therefore, we have

$$\begin{aligned} \text{Family } (S) &= \text{Family } (T \mathring{+} P) \\ &= \text{Family } (\mathscr{S}_{0,q,r} + \text{lid}) \\ &= \mathscr{C}_{q,r}, \qquad \text{using Agreement 6.} \end{aligned}$$

Note that $\chi(T) = \chi(S) - 1 = d - 1$ (say) so the formulae of Section 5.4 tell us that r is one of the numbers

$$\tfrac{1}{2}(2-d), \quad \tfrac{1}{2}(1-d), \quad -\tfrac{1}{2}d, \qquad (d = \chi(S)),$$

according as $q = 0$, 1 or 2. (For, the numbers b and c of T are 1 and $d-1$.)

We saw, in Section 5.4, that T would be orientable (i.e. $q = 0$) provided no panel in its assembly was twisted. If no panel of S is twisted, then the same holds for T, so $q = 0$; and if $q = 0$ then no panel of T is twisted, hence no panel of S is twisted either, since P is not twisted (being a lid).

These arguments therefore complete the proof of the theorem.

5.6 DEPENDENCE OF A PAPER SURFACE ON ITS ORDER OF ASSEMBLY

In this chapter, we have shown that every paper surface with boundary has a unique Gallic plan. Therefore, by the remarks on p. 56, *the Recognition Claim there is now established for any S.* Also we can now improve the Conclusion (p. 56) to:

$$\textit{If} \quad T \sim V \quad \textit{then} \quad T +_x \text{lamina} \sim V +_x \text{lamina},$$

because T and V have the same Gallic plan (which is the plan for their Family).

We must now emphasize how much the arguments of this chapter have depended on the fact that a paper surface S is assembled according to Rules 1 and 2, from panels, *in a certain order*. That is why the proof of uniqueness of S on p. 60 works: the panels of T were already assembled with an order, before P was added to form S; and T *with that order* had a unique plan because, ultimately, the very first panel has a unique order and hence the unique plan $\mathscr{S}_{0,0,0}$. However, could we not perhaps assemble S in a different way? If so, the number q might be affected. Even more, if S were repanelled in quite a different way, might not χ alter? The answers to these questions lead us to the work of the following chapter, and to a higher level of abstraction than before. But the conclusion will be that q and χ will *not* be affected by changes in order or panelling of a given paper surface S.

Exercise 5.6

1. Work out an order of assembly of the panels of $\mathscr{S}_{p,q,r}$ and of ($\mathscr{S}_{0,q,r}$ + lid).

2. Show that each panel of a paper surface can be divided up into (*a*) triangles, (*b*) quadrilaterals; and that the number χ remains unchanged. (For (*a*), see Fig. 6.2 on p. 66. For (*b*), observe that if we join a point inside a triangle to each side, then the triangle is divided into quadrilaterals.) See also Section 6.3.

3. Show that, by slightly changing the panels of a paper surface, we can ensure that only *three* panels meet at each corner. (See Exercise 6.4, No. 5, Step 1.)

6. Combinatorial invariants

In this chapter, we shall be looking at different ways in which a paper surface S might be divided up into panels. We shall show that none of these differences affects the result of our calculation of the Euler number $\chi(S)$: that is to say, we shall prove the *invariance* of $\chi(S)$. The technical problems of the proof force us to look at a wider class of paper structures than the class of paper surfaces – the class of *paper complexes*. This is one example of something that often happens in Mathematics, that one studies a special class of objects by relaxing some of the conditions that make the objects special; one sometimes appreciates the 'place' of a constraint or privilege when it is removed.

As the exercises indicate, some of the ideas in this chapter relate to an older problem, that of deciding how few colours are needed to colour a map that is drawn on a surface.

6.1 REPANELLING A PAPER SURFACE

Let S be a paper surface, assembled from panels according to the Rules 1 and 2 in the order

$$P_1, P_2, \ldots, P_n. \tag{1}$$

Suppose we found S, assembled and lying on a table, and we did not know the order in which the panels had been assembled. How then could we decide to which family S belonged?

If we examined S, we would find that it had (say) β_S boundary curves, where $\beta_S \geqslant 0$. We would also find that if e was any edge of S that was not free, then precisely two of the panels P would be joined along e.

We might be able to work out an order of assembly for the panels P, but it is likely to be different from the order (1). Our new order might involve some twists, whereas we would not immediately know whether or not the assembly (1) did. We might even prefer to 'repanel' S in a more pleasing (or at any rate different) way. Thus we might draw, upon the surface S, the boundaries of a new set of panels

$$Q_1, Q_2, \ldots, Q_m, \tag{2}$$

that lie perhaps quite askew to the original panels P. The Q's would not necessarily be the same in number as the P's, but they would fit together 'properly', i.e. if any two meet, they do so along common edges or at common corners and if e is an edge of a panel Q, and e is not free, then precisely two of the panels Q would be joined along e. Clearly, such a free edge would lie in one of the boundary curves of S. Let us call such a panelling of S a **proper** panelling. Can we now work out an order of assembly that obeys Rules 1 and 2, for such a proper panelling of Q's?

Suppose we did not know such an order of assembly. If we were trying to recognize S, we might wish to remove one of the panels, say Q, and hope to find* $S - Q$ easier to recognize. But, unless we choose Q suitably, $S - Q$ may not be a paper surface – for example, if Q was the panel C in Fig. 1.4(a). With a very complicated panelling, it might not be at all obvious which panel Q to remove, if $S - Q$ is to remain a surface.

Exercise 6.1

1. Let S be a closed surface, panelled as in (1) above. For each of the panels $P_1, \ldots, P_n$ choose a point v_i inside P_i. If P_i and P_j have an edge in common, join v_i to v_j by drawing an arc on S, within $P_i + P_j$. These arcs divide S up into a new set of panels, called the *dual* panelling of (1). Consider the Platonic polyhedra (Section 1.3) and show that if the dual panels are suitably flattened, then the duals of the tetrahedron, cube, octahedron, icosahedron and dodecahedron are respectively the tetrahedron, octahedron, cube, dodecahedron and icosahedron.

2. Show that for the dual panelling of a paper surface, the numbers of corners, edges and panels are the numbers P, E, C respectively, of the old panelling.

3. What is the relationship between the original panelling of S, and the dual of the dual panelling?

***4.** S is a closed paper surface, and each panel has exactly 3 sides. Show that $3P = 2E$, and hence that $E = 3(C - \chi)$, where $\chi = \chi_S$, while P, E, C are the numbers of panels, edges and corners. Next, show that $6(1 - \chi/C) \leqslant C - 1$ (use Exercise 2.6, No. 9). The roots of the quadratic equation $t^2 - 7t + 6\chi = 0$ are real since $\chi \leqslant 2$; show that the smaller of the roots is $\leqslant 1$ since χ is also an integer. Hence show that, since $C \geqslant 3$, then $C \geqslant \frac{1}{2}[7 + \sqrt{(49 - 24\chi)}]$. (The quadratic expression occurs in map-colouring problems; see Exercise 6.6, No. 6.)

***5.** Given the conditions of (4) above, show that if $\chi_S = 2$, then $C \geqslant 4$, $E \geqslant 6$, $P \geqslant 4$, and find analogous inequalities when $\chi_S = 1$, 0 and -2. What happens if all panels must have 4 sides instead of 3?

6.2 PAPER COMPLEXES

For the reasons stated in the last section, we shall find it convenient to allow a wider class of assemblies of panels that possibly violate Rules 1 and 2. Thus, we shall consider[T] **paper complexes**, which will be assemblies of panels that satisfy the following test:

Test

C is a paper complex if it is a sum of panels, such that any two of the panels meet (if at all) either along common edges, or at one or more common corners.

Certainly, the panels, of a properly panelled paper surface S, satisfy this test. Thus S, divided into these panels, is a paper complex. As we saw with Fig. 1.4(a), $A + B$ there is a paper complex but it is not a paper surface.

* When using such a subtraction symbol, we mean that the panel Q is removed by slicing the tape that joins Q to the rest of S; thus half the tape remains on the rest of S. (See Section 1.2.)

Also, a paper complex may fall into several quite separate pieces with no tape between them; for example, a paper complex consisting of just two panels which are not joined anywhere. On the other hand, the mode of construction of a paper surface ensures that it is always in one piece at each stage of its assembly.

Therefore, if we remove a panel Q from a paper complex C, the residue $C - Q$ is still a paper complex (perhaps in several pieces – see Fig. 6.1);

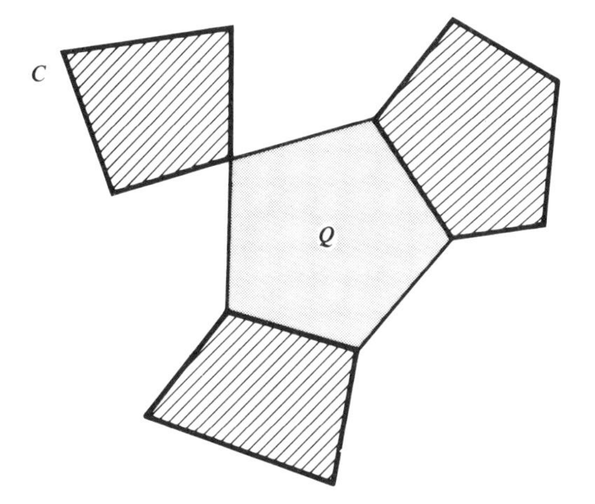

Fig. 6.1. $C - Q$ in several pieces.

this is a very convenient property, as we shall see. If now a paper complex C is composed of panels P, we can 'repanel' it by panels Q just as for a paper surface; but the panels Q must be shown to satisfy the above test, to be called a 'proper' panelling of C. Thus, for example, in Fig. 1.3(c) the first diagram violates the test at the vertex of the pyramid; the second diagram passes the test, and hence depicts a paper complex.

For two proper panellings of C, one by panels P, the other by panels Q, we can calculate *two* Euler numbers that we denote by[T] χ_P and χ_Q, just as for a paper surface. That is to say, we add the number of panels P to the total number of their corners, and subtract the number of edges, and the resulting integer is denoted by χ_P. Using the Q's similarly, we calculate χ_Q. We are going to prove:

The Invariance Theorem

$$\chi_P = \chi_Q;$$

i.e. *any two proper repanellings of a paper complex give the same Euler number.*

The proof is quite long, so we shall first give some reasons to justify devoting so much effort to it. First, the theorem tells us that the Euler number of a paper complex is an 'invariant' of the panelling – its value does not change however much we change the panelling. Thus we may choose the most convenient panelling of C in order to compute this Euler number $\chi(C)$, which is now something associated with C rather than with C-plus-panelling.

As an application let S be a paper surface such that it has b boundary curves, and Euler number c calculated using *any* convenient panelling. Then if $b+c$ is odd, the orientability number of S is $q=1$, as we saw on p. 61 Hence q is, in this case, independent of which panelling we use. (As we shall see later, the same is true when $q=0$ or 2, but more work will be needed to show that.)

Secondly, the Invariance Theorem has a famous forerunner, proved by Euler in the 18th century, in the special case that C is a single panel or a cocoon; but we have seen in earlier exercises, by using particular panellings, that

$$\chi(\text{panel}) = 1, \quad \chi(\text{cocoon}) = 2,$$

so the Invariance Theorem (and Euler) tells us that we would get these numbers for *any* proper panellings (however complicated) of a panel or cocoon respectively. That forerunner stimulated a great deal of mathematical research, to obtain analogous results for ever more complicated surfaces, solids and higher-dimensional structures. Moreover, χ began to turn up in other contexts, as for example in the theorem (that we shall prove in Chapter 8) which tells us that, for any mountainous landscape we may build on a closed paper surface S,

$$\text{peaks} - \text{passes} + \text{pits} = \chi(S),$$

i.e. we add the numbers of mountain peaks and lake bottoms, and subtract the number of mountain passes. In the last theorem, the left-hand side refers to the family of all possible smooth – that is, differentiable in the sense of calculus – height functions on S; so it (and hence χ) is a 'differential' invariant. Also, χ can be related to the curvature of a smooth surface, to vector fields[T] on it, and to other properties as well.

6.3 TRIANGULATIONS

It turns out to be technically simpler in the proof of the Invariance Theorem, if we can use only 3-sided panels. Thus, let C be a paper complex, properly panelled with panels P. If we choose a point v_P in each panel P, we can join it to each corner of P by 'spokes' (that may be curved) to divide P

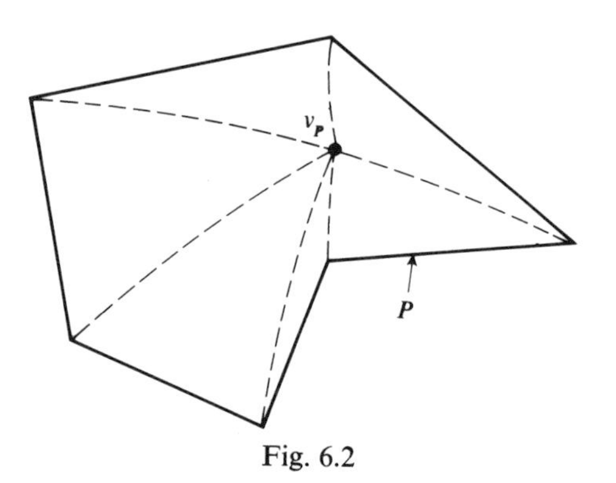

Fig. 6.2

into triangles with possibly curved sides (see Fig. 6.2). The set of all the triangles $T_1, T_2, \ldots$ we obtain in this way forms a proper panelling of C. It has an Euler number χ_T. Let us now prove a very special (and easy) case of the Invariance Theorem:

Proposition 1

$$\chi_T = \chi_P.$$

Proof. All we have to observe is that for each panel P, we introduce one new vertex v_P, and s new edges in the form of the spokes, one from v_P to each corner of P; and we replace the old single panel P by the s triangles. Thus, in the formula 'corners – edges + panels', the s new edges cancel with the s new triangles, and the new corner v_P counters the loss of the old panel P; these gains and losses therefore cancel out on each old panel, so the old number χ_P is left unchanged. Therefore $\chi_P = \chi_T$ as we claimed. This completes the proof of Proposition 1.

If a proper panelling of a paper complex consists entirely of 3-sided panels, we call the panelling simply a **triangulation**. One virtue of using triangles is this: if we remove a triangle from a lamina, the remainder is easily recognizable, provided we can make the following agreement (which in fact implies Agreement 4). For this purpose, let L be a lamina, and let* α be a curve whose ends join two corners of L, and such that α does not otherwise meet an edge of L. Suppose also that α never crosses itself. Then we want the reader to assent⊤ to an agreement which is stronger than Agreement 4:

Agreement 7

L is divided by α into two laminas M and N, such that M and N each have α as their only common edge, while

$$L = M \dot{+} N.$$

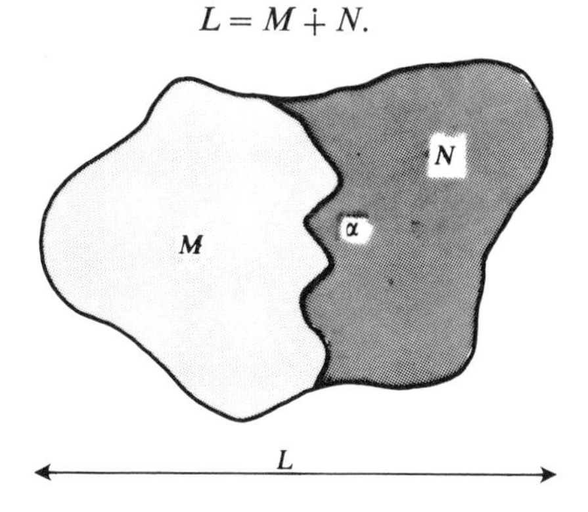

Fig. 6.3. Agreement 7 illustrated.

As a consequence, we can now prove a simple but useful result, illustrated in Fig. 6.4.

* α is the Greek letter alpha.

Proposition 2

Let L be a lamina with a triangulation, and let e be a free edge of L. Thus e is an edge of just one triangle of the triangulation. Let T be that triangle with e as one of its edges. Then L – T is either a lamina, or a pair of laminas joined at a common corner.

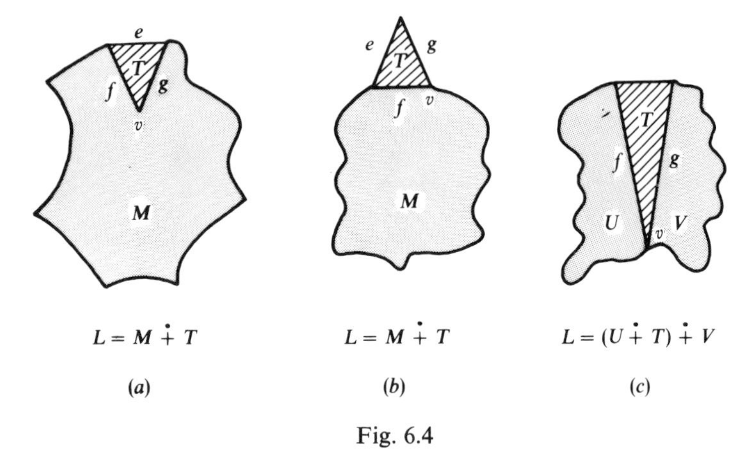

$L = M \dot{+} T$ (*a*)

$L = M \dot{+} T$ (*b*)

$L = (U \dot{+} T) \dot{+} V$ (*c*)

Fig. 6.4

Proof. The triangle T has just 2 edges besides e, say f and g, where f and g meet at a common vertex v of T. If v is not also a corner of the boundary of L, then $\alpha = f + g$ is a curve in L, to which we can apply Agreement 7, so $L = M \dot{+} T$ and $S - T$ is the lamina M, as in Fig. 6.4(*a*).

On the other hand, suppose v *is* on the boundary of L. Then the situation is either as in Fig. 6.4(*b*) or 6.4(*c*). If the first, then $L = T \dot{+} M$ and M is a lamina by Agreement 4. If the second, then g satisfies the conditions on α in Agreement 7, so $L = N + V$ (say), where N and V are laminas having g as their only common edge. But then, if $U = N - T$, we have $N = U \dot{+} T$, so by Agreement 4, U is a lamina. Clearly, U and V have only the vertex v in common. This completes the proof of the proposition.

6.4 THE INVARIANCE THEOREM FOR A LAMINA

We shall now prove the Invariance Theorem for the case when the paper complex C is as simple as possible, namely when C is a single panel, properly panelled by smaller panels P. If we regard the single panel as a panelling of C, we obtain the Euler number $\chi = 1$, since the number of edges is the number of corners, and there is just one panel, C itself. Therefore to prove the Invariance Theorem for a panel, we have to prove that $\chi_P = 1$.

Next, we split the panels P into triangles as in Section 6.3, so that C is triangulated by (say k) triangles T. By Proposition 1 in Section 6.3, we know that $\chi_P = \chi_T$, so it suffices to prove that $\chi_T = 1$.

We now use the method of mathematical induction. Let E_n – where n runs through the natural numbers 1, 2, 3, . . . – denote the proposition: 'If L is a lamina, triangulated by n triangles $A_1, A_2, \ldots, A_n$, then the Euler number χ_A equals 1.' Then E_1 is the proposition: 'If L is a lamina, triangulated by one triangle A_1, then the Euler number χ_A equals 1.' This is true since $\chi_A = 3 - 3 + 1 = 1$.

Suppose then, that we have proved propositions $E_1, E_2, \ldots, E_n$, and let us show how to deduce E_{n+1} from them. Thus we are given a lamina L, triangulated by $n + 1$ triangles $A_1, A_2, \ldots, A_{n+1}$, and we wish to calculate its Euler number χ_A. But L has a free edge e, and e lies in exactly one, say B, of the triangles $A_1, A_2, \ldots, A_{n+1}$. Therefore, by the Proposition 2, $L - B$ is either a lamina M or a pair of laminas $U + V$ joined at a common corner; so that either $L = M \dotplus T$ or $L = (U \dotplus T) \dotplus V$. Now M, U and V are laminas, triangulated by *at most n* of the triangles A. Therefore if U, for example, is triangulated by m of the triangles, then $\chi_A(U) = 1$ by the above supposition that proposition E_m has already been proved. Similarly

$$\chi_A(M) = 1 = \chi_A(V) = \chi_A(T).$$

Hence, by Exercise 2.6, No. 7, if $L = M \dotplus T$, then

$$\chi_A(L) = \chi_A(M) + \chi_A(T) - 1 = 1 + 1 - 1 = 1,$$

and the same kind of argument applies also when $L = (U \dotplus T) \dotplus V$.

This shows that if we can prove $E_1, \ldots, E_n$, then we can prove E_{n+1}. Therefore by mathematical induction, each proposition E_n is valid, for all $n = 1, 2, 3, \ldots$. In the case with which we began, we had a lamina C panelled by k triangles T, so since we now know that proposition E_k is valid, we can assert that $\chi_T(C) = 1$.

As we saw at the beginning of this section, $\chi_P(C) = \chi_T(C)$ so $\chi_P(C) = 1$, as was required. This proves the Invariance Theorem, in the case of a lamina, and is the basis for the later work.

Exercise 6.4

1. We can regard a map of Africa as a lamina L in which the different countries and lakes form a panelling of L. For economy, it is required to use only four colours to paint the panels – say red for Tanzania, blue for Kenya, etc. – in such a manner that, if two panels have a common edge, then they receive *different* colours. Can you arrange such a 'colouring'? Can you do it if blue is to be used *only* for panels representing lakes? Could you manage with only three colours?

2. Try the same problem with maps of other regions of the world, e.g. the division of England into counties, of the U.S.A. into states, etc.

3. Show that if S is a cocoon, repanelled into panels P, then $\chi_P(S) = 2$.

4. If instead S is a repanelled annulus or Moebius strip, show that $\chi_P(S) = 0$.

5. The panels of a cocoon can be regarded as an imaginary subdivision of the Earth's surface into countries, lakes and oceans (if we ignore rivers). It was guessed in the 19th century that four colours would suffice to paint the panels subject to the rule in No. 1 above – but without requiring that blue be used only for water. This 'Four-Colour Conjecture' was (correctly) proved only in 1977 by Appel and Haken, using a long computer analysis of many special cases. Their proof supersedes the weaker theorem (1890) that five colours would suffice, but the latter is provable 'by hand', as the following steps show.

Step 1. We can assume that exactly 3 countries meet at each corner v. If not, surround v by a panel Q as in Fig. 6.5 (shaded). Show that if the new map can be coloured with 5 colours, so could the old by suitably repainting Q.

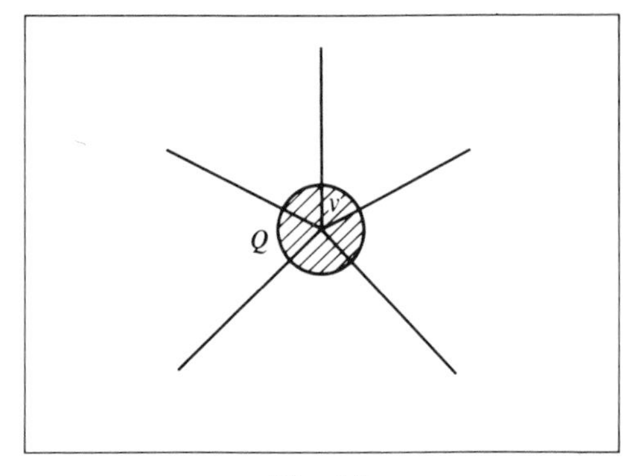

Fig. 6.5

Step 2. With the assumption of Step 1 in force from now on, show that $2E = 3C$ for the numbers E of edges, C of corners in the new panelling. Hence, since $C - E + P = 2$, show that $2P - C = 4$.
Step 3. Let p_k denote the number of panels with exactly k edges (so $p_2 = 0$). Show that $P = p_2 + p_3 + \cdots + p_k + \cdots$, while $2E = 2p_2 + 3p_3 + \cdots + kp_k + \cdots$.
Step 4. Prove that there must be at least one panel with fewer than six sides, as follows. We want to prove that one of p_3, p_4 or p_5 is not zero. If not, then in Step 3, $P = p_6 + p_7 + \cdots + p_r + \cdots$, and $2E = 6p_6 + 7p_7 + \cdots$. Use Step 2 to show that

$$4 = \left(2 - \frac{7}{3}\right)p_7 + \left(2 - \frac{8}{3}\right)p_8 + \cdots,$$

which is a contradiction since the right-hand side is negative. This contradiction arises from assuming $p_3 = 0 = p_4 = p_5$, so the opposite assumption holds, i.e. either $p_3 \neq 0$, or $p_4 \neq 0$, or $p_5 \neq 0$.
Step 5. If $p_3 \neq 0$, there is a triangular panel T, with edges e, f, g say. If e is also an edge of a panel L (see Fig. 6.6), then if we form $L + T$ as a single panel, this and the remaining panels form a panelling of the cocoon with 1 fewer panels than before. Show that if we could colour this new panelling with up to five colours, then we can still paint in T with a colour different from the colours of the panels on the other sides of the edges f and g.

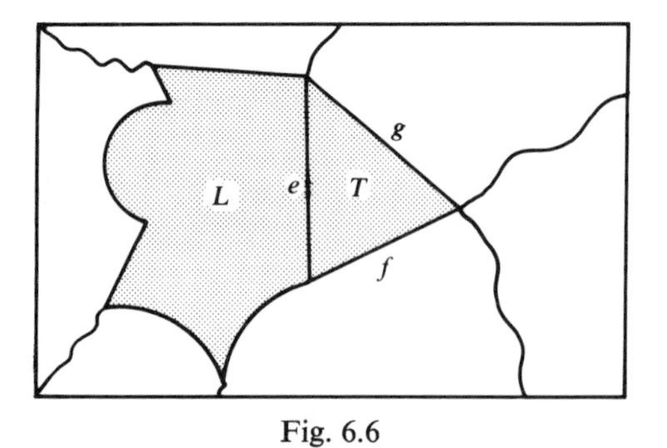

Fig. 6.6

Carry out the same argument when $p_4 \neq 0$ and when $p_5 \neq 0$, (but beware: $L + T$ could be an annulus if T had 4 or 5 sides!). For an account of the map-colouring problem see Saaty [16] and Wilson [21] for more up-to-date information and references.

6.5 REFINING A PANELLING

Let us return now to considering a general paper complex C, which has two proper panellings, by panels P, and by panels Q. We want to prove that $\chi_P(C) = \chi_Q(C)$, and by Proposition 1 in Section 6.3, we may now suppose that all the panels P and Q are 3-sided. For convenience, let us suppose that the edges of the P's and the Q's are shown in different styles, say black and dashed, while C itself is shaded. If we look at a particular one of the panels P, we see that P is itself broken up into shaded panels with dashed edges, by the edges of all those panels Q that meet P. Thus we have a 'repanelling' of P, and all its panels, R, R', R'', etc., have dashed edges except for those meeting the (black) boundary of P. Of course some of the dashed edges might well have run along the edges of P, as with R in Fig. 6.7. The panels R may have more than 3 sides; these have tape of finite width, so the number of such panels is finite.

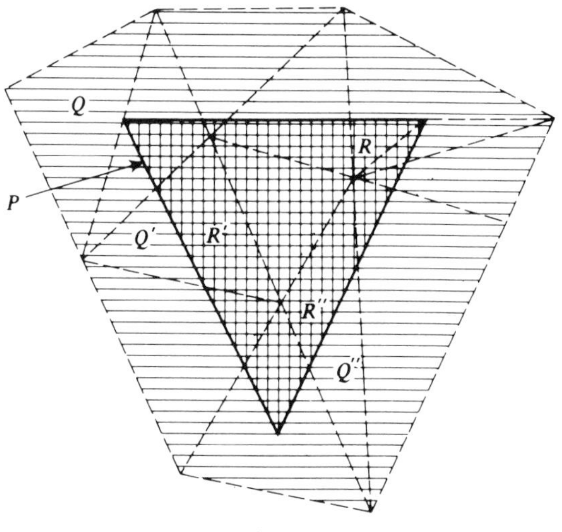

Fig. 6.7

In a similar way, every one of the panels $P_1, \ldots, P_n$ is repanelled by 'small' panels R, so these small panels taken all together, form a third, proper, panelling of the whole of S. We therefore have *three* repanellings of S, by the panels P, the panels Q, and the panels R. With each of these proper panellings we have Euler numbers that we denote by $\chi_1(C)$, $\chi_Q(C)$, $\chi_R(C)$ to show their dependence on both C and the panellings.

We now observe that the R-panelling – consisting of all the panels R – can be thought of in two ways. First, it was formed by using the Q's to repanel each separate panel of the P-panelling; but we would get the same panels R if we had instead used the P's to repanel each separate panel of the Q-panelling, since each panel R is common to one P and one Q. Also, the R-panelling is 'finer' than the P-panelling, since each panel P is repanelled by some of the R's; similarly the R-panelling is finer than the Q-panelling (or a 'refinement' of it).

Our strategy will therefore be to show that if a paper complex has two proper panellings, of which one is finer than the other, then the Euler numbers of the two panellings are equal. Once that is proved, we can conclude that

$\chi_P(C) = \chi_R(C)$ because the R-panelling refines the P-panelling,
$\chi_Q(C) = \chi_R(C)$ because the R-panelling refines the Q-panelling,

so

$$\chi_P(C) = \chi_Q(C),$$

which is what we wanted to prove. This brings us therefore to the final stage of our proof of the Invariance Theorem.

6.6 THE EULER NUMBERS OF A PANELLING AND A REFINEMENT

In this section, we wish to compare the Euler numbers $\chi_P(C)$, $\chi_R(C)$ when C is a paper complex, properly panelled by panels P, and by panels R, where the R-panelling refines the P-panelling. Using Proposition 1 of Section 6.3 again, if necessary, we may① suppose that all the P-panels are 3-sided.

Rather as we did in Section 6.4, we shall use mathematical induction. For each natural number $n = 1, 2, 3, \ldots$, let F_n denote the proposition: 'If Y is a paper complex, and $T_1, \ldots, T_n$ a triangulation of Y with just n triangles, then $\chi_T(Y) = \chi_S(Y)$, for any panelling by panels S, that is finer than the T-panelling.

Certainly F_1 holds, because if Y is a paper complex, triangulated with just one triangle T_1, then $Y = T_1$ and $\chi_T(Y) = 1$; and $\chi_S(Y) = 1$ for *any* panelling S of Y, as we proved in Section 6.4, since Y is a triangle (hence a lamina).

Let us therefore assume that $F_1, \ldots, F_n$ hold, and then deduce F_{n+1}. Thus we have a paper complex Y, triangulated by $n+1$ triangles $T_1, \ldots, T_{n+1}$ and this T-panelling has a finer panelling by panels $R_1, \ldots, R_r$. We must deduce that $\chi_T(Y) = \chi_R(Y)$.

Remove T_1 from Y. Then each panel R lies either on T_1 or on $Y - T_1$ without overlap. Therefore we have

$$\chi_R(Y) = \chi_R(Y - T_1) + \chi_R(T_1) - u, \qquad (3)$$

say, where u is the difference between the numbers of common corners and common edges of $Y - T_1$ and T_1, in the panelling R. We have a similar equation

$$\chi_T(Y) = \chi_T(Y - T_1) + \chi_T(T_1) - v, \qquad (4)$$

in which everything refers to the panelling T. Now T_1 is a triangle, so it has either 0, 1, 2, or 3 edges (and their ends) in common with $Y - T_1$; hence v is 0, 1, 1, or 0 respectively. But whenever the R-panelling introduces a vertex in one of these edges, it introduces also a new edge, and these cancel out when the difference is taken to form u. Hence $u = v$.

Now $\chi_R(T_1) = \chi_T(T_1) = 1$ by the proposition F_1. Also, $Y - T_1$ is a paper complex, triangulated by the n triangles $T_2, \ldots, T_{n+1}$; and those R-panels that lie on $Y - T_1$ form a refinement of this triangulation. Hence, by proposition E_n, we have $\chi_T(Y - T_1) = \chi_R(Y - T_1)$.

Therefore, the right-hand sides of equations (3) and (4) are equal, so $\chi_R(Y) = \chi_T(Y)$, and we have deduced the proposition F_{n+1} as a consequence of propositions $F_1, F_2, \ldots, F_n$ (although we did not explicitly need $F_2, \ldots, F_{n-1}$). By mathematical induction, then, F_n holds for all n.

We apply this to the situation we started with at the beginning of this section, where C was a paper complex, triangulated by (say k) triangles $P_1, \ldots, P_k$. Hence, by the proposition F_k,

$$\chi_P(C) = \chi_R(C).$$

At the end of the last section we saw that this last equation is all that is needed to be able to complete the proof of the Invariance Theorem, so we have finished the entire proof.

Exercise 6.6

1. In the last part of the proof, we remarked that we did not explicitly need $F_2, \ldots, F_{n-1}$, but only F_1 and F_n, to deduce F_{n+1}. What was the technical reason for not needing the others?

2. If S is the real projective plane (see p. 22) show that $\chi(S) = 1$; while if S is a torus or Klein bottle, then $\chi(S) = 0$. (We now consider S with any panelling.)

3. Show that the arguments in Exercise 6.4, No. 5, apply when the cocoon is replaced by a real projective plane S, to show that there will be a panel in S with fewer than six sides. In Step 5 there, however, one now needs *six* colours to carry it through.

4. In Fig. 6.8, we show a panelling of the Moebius band, with a suggested colouring by six colours indicated by the numbers. Show that this panelling cannot be coloured with fewer than 6 colours. Hence show that the real projective plane has a panelling which cannot be coloured with fewer than six colours.

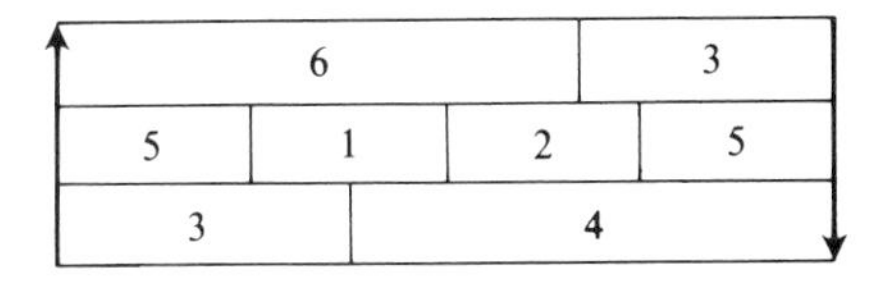

Fig. 6.8

5. How many colours are needed for colouring a panelling of an annulus?

***6.** Prove Heawood's Theorem: *If the surface S is closed, with P panels and Euler number χ, then if $\chi < 2$, it suffices to use N colours, where*

$$N \leqslant \tfrac{1}{2}[7 + \sqrt{(49 - 24\chi)}].$$

This gives the correct numbers 6 for the projective plane (where $\chi = 1$) and 7 for the torus (where $\chi = 0$), as well as the conjectured number 4 for a cocoon ($\chi = 2$); but the proof of Heawood's Theorem does not work in these cases. It can also be shown (and is much harder) that every closed surface has a panelling that requires the full number of colours. For details see Coxeter [6], p. 394.

7. Let v be a corner of a paper surface X triangulated by n triangles. If vab, vbc are neighbouring triangles in X, show that we can replace them by two triangles of which only one has v as a corner. Hence show that we can change the triangulation of X so that it still has n triangles, but no more than 3 have v as a corner. [Exchange the edge vb for a new edge ac, thus simplifying X at v, at the cost of complicating other corners.]
8. Let X be a paper surface with boundary B. Suppose there is a panel P with a corner v such that v lies in B but neither of the two edges av, bv of P (through v) lies in B. Call such a corner v a 'bad' corner. Show that it is possible to alter all the panels at v slightly, so that v is no longer a 'bad' corner in the resulting repanelling, but without altering the total number of panels, or introducing further 'bad' corners. (A solution is indicated in Fig. 7.2 in the next chapter.)
9. Show that if a paper surface X has a panelling such that no panel P has a 'bad' corner in the sense of No. 8, then each connected piece Y of $X-P$ is properly panelled (in the sense of p. 64) by the panels of X that it contains; in particular, verify that the boundaries of Y consist of Jordan curves without crossing points.
10. In No. 9, let g_P denote the maximum number of panels in any connected piece of $X-P$ when the panel P is removed from X. Thus, if X has n panels, then $g_P \leqslant n-1$. Show that if $g_P < n-1$, then $X-P$ has at least two connected pieces Y, Z, of which Y (say) has g_P panels, and Z has a panel Q with $g_Q > g_P$. Hence show that if we choose R to be a panel with g_R as large as possible, then $g_R = n-1$ and $X-R$ is a single piece.
11. Find triangulations for various surfaces, using the least possible number of triangles. [Use Exercise 6.1, Nos. 4 and 5, and observe that P is least (for a given χ) when E and C are; also the least number of triangles for a sphere is 4, given by a tetrahedron.]

7. Order of assembly, and orientability

In this chapter, we shall consider two questions that were suggested by the work of Chapter 5. The first was this: is the orientability number q of a paper surface S an invariant of the panelling? We have already seen (p. 66) that if $q = 1$, then this number is indeed an invariant, because β and $\chi(S)$ are invariants. A procedure is now needed for separating the cases when q is even. Our conclusion will be recorded in the Orientability Theorem and its consequences (Section 7.3).

Our work will be made easier if we first deal with the question mentioned at the end of Chapter 5: if a paper surface S has a P-panelling, can we work out a numbering of the panels P such that they can be assembled *in that order*, subject to Rules 1 and 2? If we can find such an order of assembly, we call the P-panelling (with that numbering) an *ordered assembly* of S. Although we cannot always find such an order, we can always do the next best thing, which is quite useful, and recorded in the Assembly Theorem (see Section 7.2). When we do not know the order of a panelling, we call the surface a 'mystery surface' and describe these more carefully as follows.

7.1 MYSTERY SURFACES

The importance of having an ordered assembly for S, of panels Q_1, Q_2, . . ., Q_m, is that $S - Q_m$ is a paper surface with its own ordered assembly Q_1, Q_2, . . ., Q_{m-1}. Hence, as we saw in the proof of the Fundamental Theorem (Section 5.2), we can decide on Family(S) once we know Family ($S - Q_m$) and the manner in which Q_m is added (as ear, bridge, etc.).

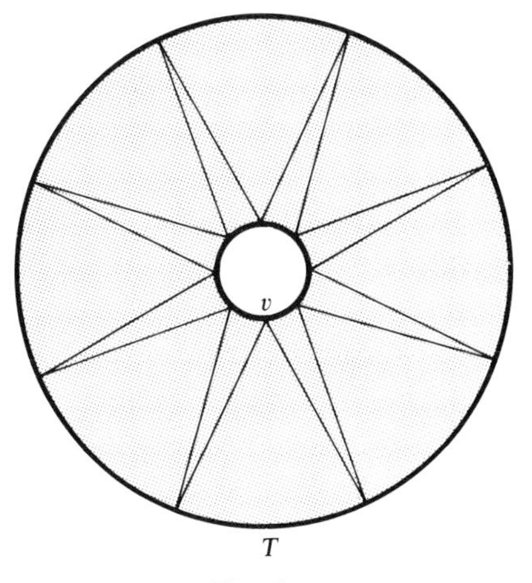

Fig. 7.1

Not every panelling can be given an order of assembly, as described above. For example,Ⓣ[a] Fig. 7.1 shows a triangulation of an annulus A, which is such that if *any* triangle T is removed, then* $A - T$ is only a paper complex, not a paper surface. Thus no triangle could be the last in any ordering, so no such ordering exists.

Nevertheless, this example is no real set-back, because our reason for wanting to turn a panelling of S into an ordered assembly was for the purpose of being able to recognize Family (S), when we might have lost the original order of assembly of S. This purpose can still be achieved, because we can prove a result (the Assembly Theorem, in Section 7.2) which is satisfactory for the purpose. To make matters quite clear, let us use a special (but temporary) name: a *mystery surface* will be any paper complex W such that (i) no edge belongs to more than two of the panels, andⓉ[b] (ii) the free edges form boundary curves without crossings. For example, $A - T$ in Fig. 7.1 satisfies condition (i), but not condition (ii) at the corner v. This allows for the possibility (which we can imagine but which we shall show not to be a real one) that there might exist objects that look like panelled paper surfaces, but which could never have been put together with an order of assembly that used any possible panelling. Of course, it still makes sense to repanel or triangulate a mystery surface, as we did for a paper surface. To exclude one possible difference between mystery surfaces and paper surfaces, we must insist that W is in one single piece (the technical name is 'connected'). That is to say, if P_a and P_b are any two of the panels of W, then we can pass from P_a to P_b along some chain of the panels P, such that each has an edge in common with the next. A paper surface is always connected, since at each stage we always join a new panel to the earlier ones.

Exercise 7.1

1. Show that W is connected if there is *one* panel that can be joined to any other, like P_a to P_b above.

2. Show that every regular polyhedron (Section 1.3) is connected.

3. Show that the plans $\mathscr{S}_{p,q,r}$ are all connected.

4. In a paper complex C, let v be a corner, and let D consist of all panels of C that have v as a corner. Show that D is a paper complex. If also C is a mystery surface, need D be a lamina? (D is called the 'Star' of v.)

5. If C is a paper complex such that the Star of each corner is a lamina, show that C is a mystery surface.

7.2 THE ASSEMBLY THEOREM

The result we shall prove is as follows, and tells us at once that mystery surfaces are really ordinary paper surfaces after all.

The Assembly Theorem

Let W be a connected, mystery surface. Then it can be repanelled so as to be given an ordered assembly. (Hence W is in fact a paper surface.)

* See footnote on p. 64.

Proof. We use mathematical induction on the number of panels in W.

Let H_n denote the proposition: 'If S is a connected mystery surface with a panelling by n panels $T_1, \ldots, T_n$, then S can be repanelled so as to be given an ordered assembly which also has n panels.'

Certainly H_1 holds, the assembly instruction being 'take T_1: that is S'. Suppose then that we have proved $H_1, H_2, \ldots, H_n$, and let S be a paper surface with a panelling by $n+1$ panels $T_1, \ldots, T_{n+1}$. Let us remove T_j to leave a paper complex C, so $S = C + T_j$. If C is a mystery surface, it is properly panelled by n panels T_i $(i \neq j)$, and therefore by proposition H_n it has an assembly into panels $P_1, P_2, \ldots, P_n$. In that case, $P_1, P_2, \ldots, P_n$, T_j (in that order) form an assembly for S, if appropriate new corners are added on the edges of T_j.

Now C will be a mystery surface if T_j is a lid for it (so no edge of T_j is free) or if $S = C \dotplus T_j$. To ensure that we can find a panel T_j such that C is a mystery surface (if no panel forms a lid) we may need to modify the panelling as follows. If no panel forms a lid, then S has a boundary B consisting of one or more Jordan curves. It might happen that B has a 'bad' corner v, i.e. some panel T has v as a corner, but neither of the two edges av, vb of T (through v) lies in B. But since S is a mystery surface, exactly two edges pv, vq of B pass through v. Let r, x, s, y be the points shown in Fig. 7.2, wherein the rays from v indicate the edges of all panels

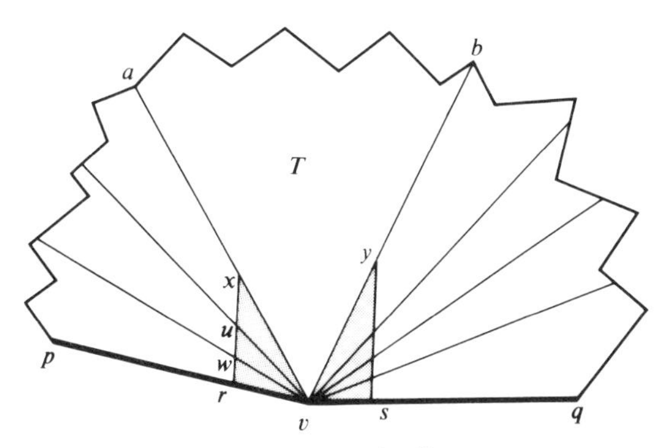

Fig. 7.2. Eliminating the 'bad' corner v.

having v as a corner; and the lines rx, sy are short enough so as not to cut any other edges of those panels. Now add the (shaded) triangles vxr, vys to T to form a new panel, thus removing a small triangle such as uvw from each panel (other than T) that had v as a corner. This process repanels W so that the number of 'bad' corners is reduced by 1 without changing the total number, $n+1$, of panels. By repeating the process sufficiently often, we can eliminate all 'bad' corners, and still have $n+1$ panels in all. Suppose this elimination is completed, leaving a modified panelling $R_1, R_2, \ldots, R_{n+1}$ of S.

Thus, for any panel E of this R-panelling, E is not a lid for $S - E$, and $S - E$ is not just a paper complex; it may be divided by E into two separate pieces, or it may be in one piece, but in either case, each piece forms a paper *surface*, because it is properly panelled by the R-panelling and its boundary curves have no crossing points.

We claim that we can always find a panel E for which $S - E$ consists of a single piece. For, we follow the prescription of Exercise 6.6, No. 10, and choose[Ⓣa] a panel E for which the maximum number g_P of panels in any piece of $S - E$ is as large as possible (and then it turns out, as in the exercise, that $S - E$ is connected). Let then such a panel E be chosen. Therefore, applying proposition H_n to $S - E$ in the modified panelling, we can assert that $S - E$ has an ordered assembly into panels $Q_1, Q_2, \ldots, Q_n$, say. Therefore, adding on E as the last panel, we obtain the assembly $Q_1, Q_2, \ldots, Q_n, E$, for S itself. (Note that $S - E$ may have a bad corner as a result of removing E from S, so it may be necessary to modify its panels as we modified the T's, before we can get the Q's.)

This shows that H_{n+1} follows if $H_1, H_2, \ldots, H_n$ are already proved. By mathematical induction, therefore, H_n holds for all n. Thus every mystery surface has an assembly, and this completes the proof of the theorem.

7.3 THE ORIENTABILITY THEOREM

Having added the Assembly Theorem to our armoury, we can think of attacking the problem of the orientability number q, which began this chapter. If, as we might guess, the orientability number really is an invariant, then we need to look for a geometrical property that expresses that fact. Not surprisingly, therefore, we now go back to the kind of orientability as we understood it in Section 3.5.

When we discussed it there, we were concerned with the assignment of orienting arrows to the panels P of a paper surface S, in such a way that the Internal and External Properties (p. 28) were satisfied. We said that S would be called orientable if we could find such an assignment of arrows. But suppose that we repanelled S with a proper* Q-panelling. Could we *then* assign arrows to the panels Q, so as to satisfy the Internal and External Properties? If we could, we shall say that the Q-panelling *itself* is orientable.

Our program will now be to prove what your attempts at Exercise 3.5 may already have led you to guess[Ⓣb]:

The Orientability Theorem

Suppose the paper surface S has a proper panelling which is orientable. Then every proper panelling of S is orientable.

Once this theorem is proved, we can conclude at once that if we can find a single panelling of a paper surface that we can *prove* to be not orientable, then no panelling of the surface is orientable. An all-or-nothing affair! It then becomes legitimate to call S orientable or non-orientable as the case may be, since the theorem tells us that these properties are *invariants of the panelling*.

Exercise 7.3

1. Prove that the plan $\mathscr{S}_{3,0,2}$ is orientable (by assigning orienting arrows) and hence prove that $\mathscr{S}_{3,1,2}$, $\mathscr{S}_{3,2,2}$, and $\mathscr{S}_{3,7,2}$ are non-orientable.

* In the sense of p. 63. We shall use the term 'proper panelling' when we are prepared to consider a panelling which is not necessarily an ordered assembly.

2. Prove that $\mathscr{S}_{p,0,r}$ is orientable, and hence that $\mathscr{S}_{p,q,r}$ is non-orientable if $q \neq 0$.

3. Explain why the Orientability Theorem, once proved, will show that the orientability number is an invariant of the panelling. (For an explanation, see p. 81.)

7.4 PROOF OF THE ORIENTABILITY THEOREM

The proof of the Orientability Theorem parallels that[T] of the Invariance Theorem in Chapter 6.

Step 1. Thus, if we divide up the panels P of S into triangles T as we did in Section 6.3, we see that the T-panelling is orientable if the P-panelling is, and conversely (i.e. the P-panelling is orientable if, and only if, the T-panelling is orientable); see Fig. 7.3.

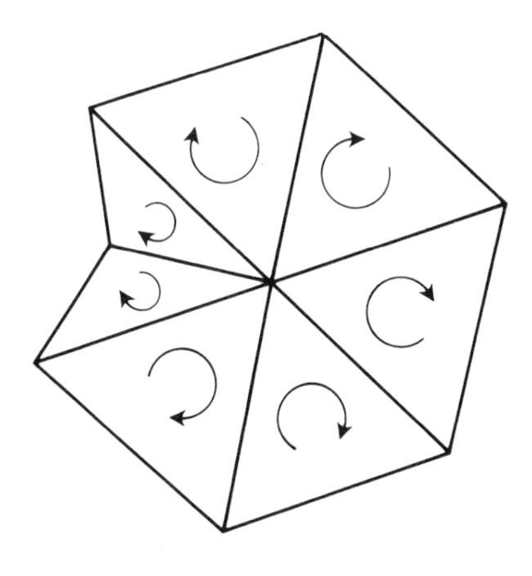

Fig. 7.3

Step 2. Next, copying Section 6.4, we show that if C is a panel, then *any* triangulation of it is orientable. This is clear from Fig. 6.4, because if we orient C with a single arrow t, and use t to orient the T-triangulations of M, U and V there, then we can orient the whole triangulation of C. (For example, use t to orient U; this gives a direction in which to orient the single triangle T, and this in turn gives the direction in which to orient V.) By thus making the directions of the new arrows agree with t on the boundary of C, we have 'induced' them by t. Conversely, an orientation of the T-triangulation of C induces a single arrow round the boundary of C – by the External Property (p. 28).

Step 3. Now let S be any paper surface, with a proper P-panelling and a proper Q-panelling. We saw that we can construct an R-panelling which refines each of the other two. Our strategy is therefore to show that if S has two proper panellings, one finer than the other, then each of the panellings is orientable if the other is. Once this is proved, then we shall have at once: if the P-panelling is orientable, so is its refinement the R-panelling; hence so is the Q-panelling because its refinement is orientable.

Step 4. Let G denote the proposition: 'If Y is a paper surface, and $T_1, \ldots, T_n$ a triangulation of Y with just n triangles which has a refinement by an R-panelling, then the triangulation is orientable if, and only if, the R-panelling is orientable.' (We do not use induction on n here.)

Suppose then that the T-triangulation is orientable so that each triangle T_i has been assigned an orienting arrow t_i in such a way that the Internal and External Properties (p. 28) are satisfied. Now T_i is panelled by some of the panels of the whole R-panelling; so by Step 2, if R_j is a panel of the R-panelling that lies in T_i, it acquires an orienting arrow r_j induced by t_i. We claim that the arrows r_j form an orientation of the R-panelling of S. To justify this we must show that the arrows r_j satisfy the Internal and External Properties. For the Internal Property, let e be a common edge of panels R_j, R_k; if these both lie in the same panel T_i then their arrows are opposed on e by Step 2, while if they lie in different panels T_i, T_q then these have a common edge d that contains e. On e, the arrows t_i and t_q are opposed, since they satisfy the Internal Property; and they agree with r_i and r_j, so the latter are opposed on d as required. For the External Property the directions of arrows r_j along any boundary curve J of S, cannot be opposed, because (by Step 2) they agree with arrows t_i which themselves satisfy the External Property. Therefore every one of the R-panels is assigned an orienting arrow, so that the Internal and External Properties hold. Hence the R-panelling is orientable.

If, instead, it had been the R-panelling of Y that was orientable, we could conclude that the T-triangulation was orientable, by interchanging R and T in the last bit of the argument. (That bit uses the 'T only if R' part of the proposition G above; the interchange uses the 'T if R' part.)

As a consequence, if S is a paper surface with a triangulation and a refinement of it, then the triangulation is orientable if, and only if, the refinement is orientable.

Step 5. Finally we tidy the logic. If S is a paper surface with a proper P-panelling and a proper Q-panelling, suppose that the first is orientable. We form a T'-triangulation of the P-panelling as in Step 1, and a T''-triangulation of the Q-panelling, similarly. These triangulations have a refinement by an R-panelling, as in Step 3. We then pass along the chain $P \to T' \to R \to T'' \to Q$, to use the orientation of the P-panelling to orient the Q-panelling, via the stages shown by the arrows. On the way we use Step 1, Step 4, Step 4, Step 1 respectively.

This completes the entire proof of the Orientability Theorem.

7.5 CALCULATION OF THE ORIENTABILITY NUMBER

The Orientability Theorem can now be applied to any proper P-panelling of a paper surface S. In Exercise 3.5, No. 6 we saw that if S is orientable, and we add a panel A as an ear or bridge (neither being twisted), then $S + A$ is still orientable.

If A has a twist, then $S + A$ is not orientable, because if we add A to the P-panelling, the resulting panelling cannot be oriented, for the same reason as with Fig. 3.5(*b*). Hence, by the Orientability Theorem, *no* panelling of $S + A$ is orientable, so $S + A$ is not orientable.

We saw in Exercise 4.4A, No. 6, that $\mathscr{S}_{p,0,r}$ (in its 'simple' panelling) is orientable; hence by the Orientability Theorem, $\mathscr{S}_{p,0,r}$ is orientable with any panelling.

The previous remarks about $S+A$ then show at once that $\mathscr{S}_{p,q,r}$ is not orientable if $q \neq 0$; they show too, by Agreements 1 to 6, that if S is orientable, so is every paper surface in Family (S). Therefore, to work out Family (S), we proceed as follows. Suppose first that S is not closed, i.e. $\beta(S) \neq 0$. If S is given with a panelling whose order of assembly we know, then we can use the method suggested on p. 55. However, if all we know is a proper panelling of S, then we can find a repanelling by panels P and an order of assembly, by the Assembly Theorem. Using the inductive method of p. 60 we then find a plan $\mathscr{S}_{p,q,r}$ for S with the repanelling P, such that $q=0$, 1 or 2. If $q=0$ then the repanelling P can be oriented; therefore so could the original panelling of S, by the Orientability Theorem. By the same theorem, if $q=1$ or 2, then no panelling of S can be oriented. This decides for us whether or not S is orientable; and then we know p, q and r, by the calculations of Section 5.3 (p. 59), in terms of $\beta(S)$ and $\chi(S)$, which numbers can be calculated from *any* convenient repanelling of S. We have proved, then, that *all three numbers are invariants of S*, and independent of any proper panelling of S. Further, we have identified S in terms of the Gallic plan $\mathscr{S}_{p,q,r}$ which is the typical representative of Family (S).

If, instead, S is closed, we first remove a convenient panel, to leave a surface T. We then identify T as above by finding its plan $\mathscr{S}_{p,q,r}$ (where $p=0$). Hence Family (S) is $\mathscr{C}_{q,r}$, and we are finished.

Exercise 7.5

1. Convert the procedure described above, for deciding Family (S), into a flow chart. (You may wish to elaborate this, depending on your knowledge of computer programming.)

2. Show that it is possible to give a rule that says what a 'boundary curve' of a paper complex C is. Hence, show that we could have proved the Orientability Theorem without needing the Assembly Theorem. [Hint: At a corner X of Q in Fig. 6.1, a free edge e, that meets X, is an edge of just one panel P. This panel has just one other edge f meeting X. Hence the 'boundary' of C must include e and f in that order, before moving from P onto another edge at X.]

3. Write a flow chart for deciding Family (S) when S is a mystery surface. Your procedure should begin by modifying the panels to eliminate 'bad' corners, as on p. 77; and it should then give a process for picking the panel E such that $S-E$ is a connected mystery surface. It can conclude with the flow chart of No. 1 above. (Writing the early portion can help you to understand the inductive processes involved in the proof of the Assembly Theorem, in Section 7.2.)

4. Settle the point raised on p. 46, where we asserted that $\mathscr{S}_{0,2,0}$ and $\mathscr{S}_{0,0,1}$ are in different families. [One is orientable, the other not; but if they were in the same family, then they would each have a prescription of the form (13) on p. 55, the same for each surface. This shows that the two twists in $\mathscr{S}_{0,2,0}$ cannot be removed using only the changes envisaged in Agreements 1–7, so a more drastic type of change would be needed. If, however, we allow a panel to be cut across, and glued up differently, then $\mathscr{S}_{p,q,r}$ could be converted into a planar region with $p+q+2r$ holes, and we would not have a particularly satisfying theory.]

8. Morse Theory of a paper surface

8.1 THE MOUNTAINEER'S EQUATION AND PHYSICAL GEOGRAPHY

In this chapter, we shall describe the main ideas that lead to the proof of the equation we mentioned in passing, in Section 6.2, that for a paper surface S which possesses a 'mountainous landscape', the alternating sum that gives $\chi(S)$ is related to another alternating sum:

The mountaineer's equation

$$\textit{pits} - \textit{passes} + \textit{peaks} = \chi(S).$$

This was known to the great German mathematician Kronecker in the 19th century, and its modern development has had a great influence on the subject of Topology (which grew from the study of surfaces), due to the American mathematician Marston Morse – a relative of S. F. B. Morse, who invented the Morse Code. The theory is therefore called 'Morse Theory'.

Let us first consider the mountaineer's equation as it applies to the physical geography of the Earth (so S is now a sphere, in the cocoon family). We shall follow the approach of the Scottish mathematical physicist, James Clerk Maxwell, who wrote a paper entitled 'On Hills and Dales' in 1869, to be found in Vol. 2 of his *Collected Works*, p. 233; he begins by apologizing for possibly repeating ideas of Cayley from 15 years earlier.

The words 'pits' and 'peaks' stand for the numbers of valley bottoms and mountain tops on Earth, and such bottoms and tops are relative to our way of assigning a height to each point of Earth's surface, above or below sea-level. We have a peak at any point that is higher than all other points in the neighbourhood, and a pit is lower than all points in *its* neighbourhood. We assume for simplicity that there are no plateaus or volcano rims; or rather, that none of these natural occurrences is completely horizontal, but each has a highest and a lowest point. This way of measuring height is not as simple as it sounds (as Maxwell points out) but we ignore the practical difficulties here. However, we might instead take a scale-model globe that is mounted on a stand, rest it on a table-top and measure all heights from the table. This would then qualify certain mountain tops in Australia for being *pits* relative to this new method of measuring heights, and the new counts of 'peaks' and 'passes' might be also different from the old. Nevertheless, if we counted them and inserted the numbers in the mountaineer's equation, the mathematical proof tells us that we shall still get 2 (= χ(sphere)) for the alternating sum. This illustrates the important fact that the mountaineer's equation is a statement about both the surface

S and the way in which we choose to measure heights: for the same S, two different ideas of height may yield differing numbers of pits, peaks and passes, but nevertheless the alternating sums come to the same answer in each case (namely $\chi(S)$).

When S is the surface of the Earth, and we use height above sea-level, we can see why the mountaineer's equation should hold in the following way, using a slightly more picturesque version of Maxwell's argument than he did. First imagine that the Earth is completely dry of water, but otherwise keeps its topography of depressions, etc. Now let it rain incessantly, as in Noah's Flood, until the whole of Earth lies under water. Suppose we (or God) can observe the shape of the entire water-level, as the rain falls and the level rises. The water-level is always a planar region until the last patch of dry ground disappears, but its Family changes as the water rises. At first, the rain forms separate puddles as it collects in the pits, and we suppose that it soaks out of mountain hollows as long as they are higher than the general water-level. Eventually, a pair of puddles will join into one, as the water rises to form an 'isthmus' connecting them. On the other hand, a puddle might gradually increase in size until it completely surrounds an island, to form a moat. The ring of the moat closes as the water rises, to form a 'bar', and the number of moats is exactly the number of bars.

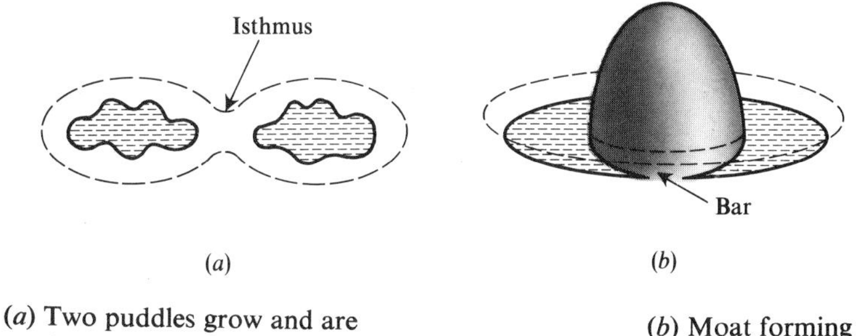

(*a*) Two puddles grow and are joined by an isthmus.

(*b*) Moat forming across a bar.

Fig. 8.1

Suppose there are p pits. Then during the whole Flood, p puddles will be started, and the number I of isthmuses needed to join them all up to make the final single sheet of water is therefore given by the equation

$$p = I + 1. \tag{1}$$

Next, suppose that B moats are observed to form. Each moat has two boundaries, and we need one peak to fill the inner one; this peak is the last point of the enclosed island to drown under the rain. When all such inner boundaries have disappeared, there is still one last outer one to be closed up as the rain finally engulfs Mount Everest. Therefore the number P of peaks satisfies the equation

$$P = B + 1. \tag{2}$$

Adding equations (1) and (2) we get

$$P + p = (B + 1) + (I + 1),$$

so if we denote $B + I$ by s we get

$$p - s + P = 2 = \chi(\text{sphere}). \tag{3}$$

Recall that B was the number of bars (as well as the number of moats). The lie of the land in the neighbourhood of a bar or isthmus is exactly that of what is called a pass (a term which ordinary speech associates with mountainous country), so equation (3) is the mountaineer's equation for the Earth when s is taken to be the count of passes. In the more detailed study below, we shall see that bars and isthmuses really are passes, using a different mode of recognition.

Turning from this special case we shall now consider a proof of the mountaineer's equation for a general paper surface instead of the surface of the Earth. For simplicity, we are going to use a special way of measuring height, but we emphasize that the equation holds for all other ways of measuring, also. We begin by first saying what we shall now mean by pits, peaks and passes, and we need to be more exacting than we were with Earth, because we cannot use the descriptive device of a flood in quite the same way.

Exercise 8.1

What exactly would you mean by 'sea-level' on the Earth? Remember that the Moon makes the sea rise and fall. (For a discussion, see the quoted paper of Maxwell.)

8.2 THE MOUNTAINEER'S EQUATION FOR A GENERAL SURFACE

To keep the technicalities as simple as possible, we shall now understand the left-hand side of the mountaineer's equation in the following way. We suppose that S is placed on a flat table in such a manner that each boundary curve is horizontal. Then for each point X on S we let $h(X)$ denote the

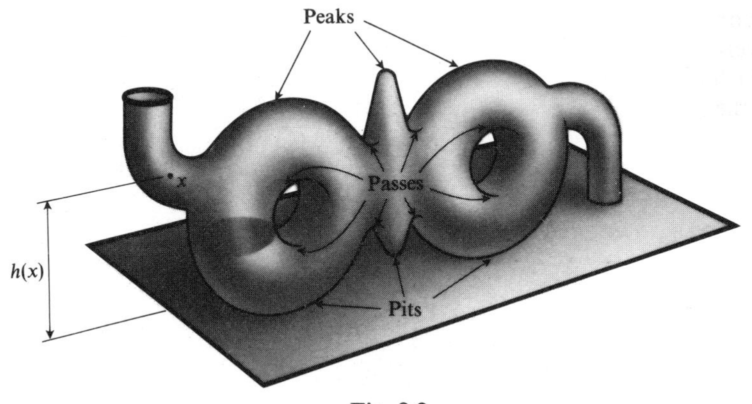

Fig. 8.2.

height of X above the horizontal. Thus, if X moves on any boundary curve, $h(X)$ remains constant. Then the pits, peaks and passes are to be understood respectively as points of S at which the function[T] h has a local minimum, maximum, or a saddle point. We shall be more informative about these 'critical points' in a moment, but it must be understood that we are supposing that there are no plateaus, volcano rims or ridges on S,

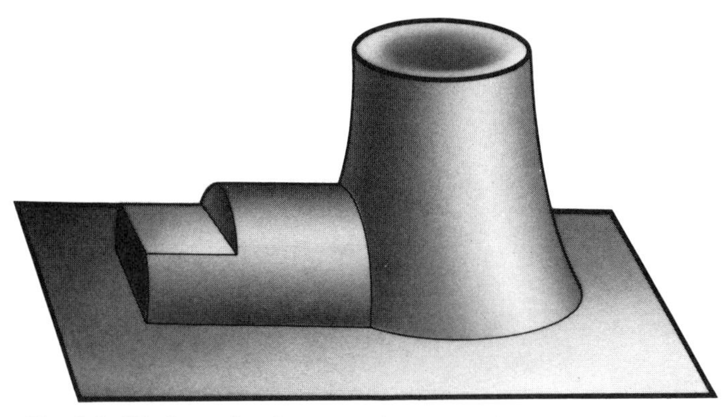

Fig. 8.3. Horizontal volcanoes, plateaus and ridges not allowed.

i.e. no panels or curves wholly composed of maxima, or of the other types of critical point. (A boundary curve does not count as such a rim.) Such plateaus and curves can always be eliminated by tilting the table. The boundary curves can then be stretched a little so as to keep them horizontal. They are never to be counted among the critical points. After this preliminary tilting of the table, we shall say that S is *properly arranged.*

Having said what we are *not* including as critical points, let us now say what we *are* counting. If we take a horizontal slice through a properly arranged surface S, or fill it to a certain depth with water, the edge of the slice (or the 'tide-mark' of the water) would consist of all points X of S at the height (say d) of the slice (or tide-mark). That is, if X is such a point, then the height $h(X)$ equals d. We call this tide-mark of points a *level* of h; it consists of the 'level d' of all points of S that have height d. If we mark levels at (say) centimetre intervals on S, the result looks like a map, on which the levels are contour lines. (See also the wood grain in the sculpture in the photograph on p.88.)

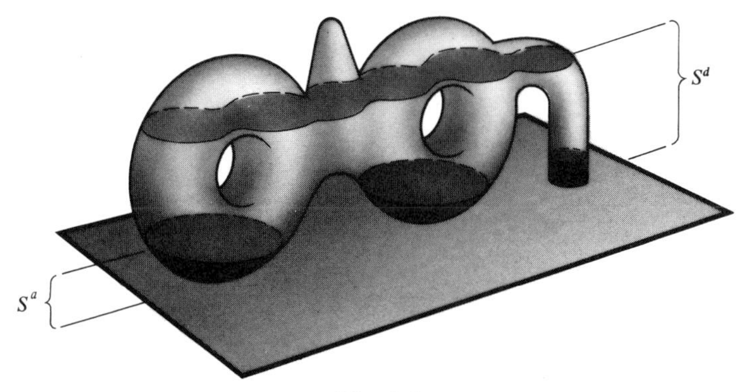

Fig. 8.4

The portion of S that lies at or below the level d will be denoted by S^d. For most of the levels d, S^d will consist of one or more paper surfaces, whose boundary curves compose the level d. For example, in Fig. 8.4, S^d is a single paper surface, while the level S^a consists of three paper surfaces (2 panels and 1 annulus).

However, for a few levels b, S^d is only a paper *complex*, not a paper surface, because one or more of the curves of level b cross themselves. For example, in Fig. 8.5, we show five such levels, b, c, d, e and f. The cross-

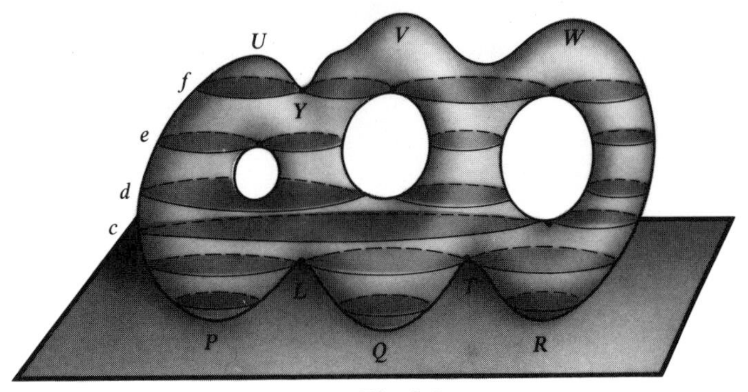

Fig. 8.5

over points are called 'saddle points'; and the level b has two such points, while c, d and e have one each, and the level f has three. The contour lines near a saddle point are arranged as in Fig. 8.6 (when we gaze down upon them). They are drawn as smooth curves, but a contour line may well have angles in it, like a polygon, if the panels of S make angles with each other.

However, the contours near a saddle point (like Y in Fig. 8.5) look like those at a mountain pass on a geographical contour map, or near the isthmuses and bars mentioned in Section 8.1. Hence the name 'pass' in the mountaineer's equation.

The contour lines near the points P, Q, R, U, V, W in Fig. 8.5 are all arranged as in Fig. 8.7; they are all closed Jordan curves, roughly concen-

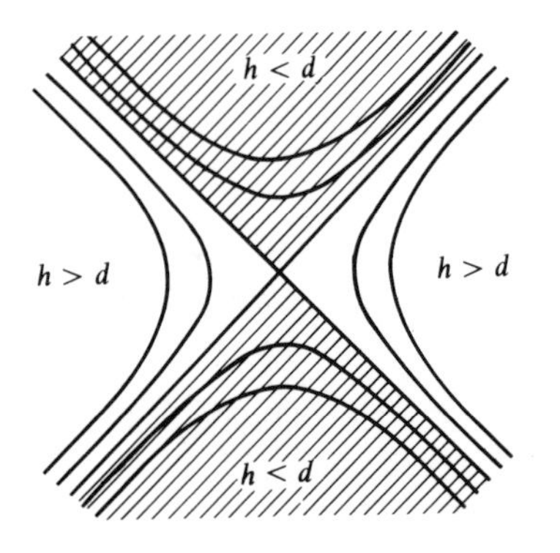

Fig. 8.6. Contours at a saddle point.

Reclining Figure, 1945–6, 75 in. The grain of the elm-wood displays the contours of a 'height'-function that measures *not* the height above the table but the age of the wood when it was formed in the tree. (Photograph by courtesy of Henry Moore, O.M., C.H.)

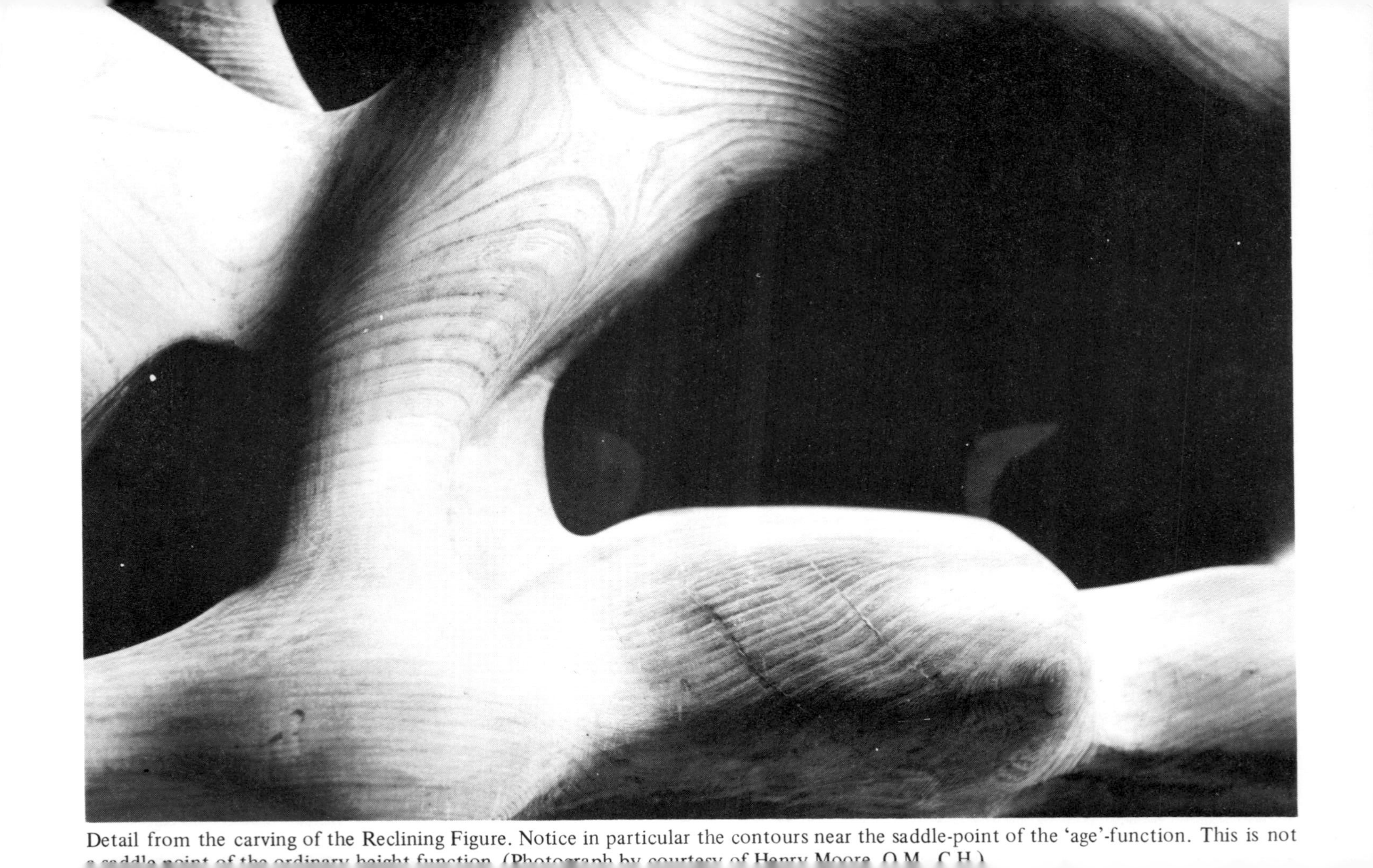

Detail from the carving of the Reclining Figure. Notice in particular the contours near the saddle-point of the 'age'-function. This is not a saddle-point of the ordinary height function. (Photograph by courtesy of Henry Moore, O.M., C.H.)

tric about a centre. At the points P, Q, R, the height increases *from* the centre, while at U, V, W, the height increases *to* the centre. Points of the

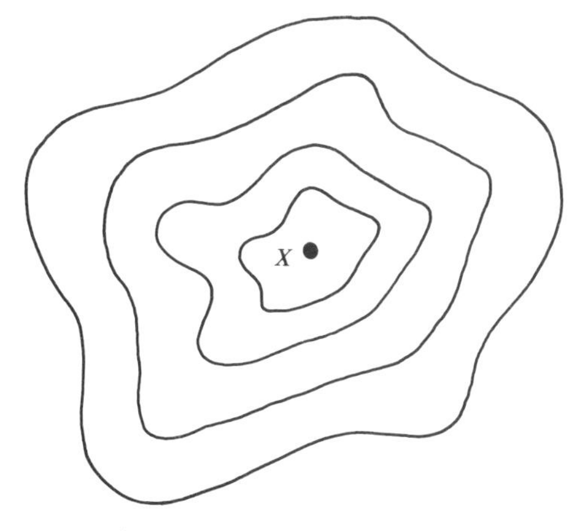

Fig. 8.7. Contours at a peak or pit.

first kind are called 'relative minima' or pits (because the centre is lower than its neighbours) and those of the second kind are called 'relative maxima' or peaks (because the centre is higher than its neighbours). The highest peak (or peaks if they are level) are 'absolute maxima' – e.g. V, W, in Fig. 8.5; similarly with the absolute minima. In our case, the absolute minima are at height zero, so in Fig. 8.5, the points of S^0 are just P, Q and R. If the absolute maxima V, W are at height H (say) then S^H is the whole of S.

The pits, peaks and passes of S, in this sense, are called the *critical points*. Any level upon which there lies some critical point, or boundary curve of S, is called a *critical level*, while all the other levels are said to be *ordinary*. Clearly,Ⓣ a paper surface has only a finite number of critical levels, say $d_0, d_1, d_2, \ldots, d_{n+1}$, in increasing order of heights, where $d_0 = 0$ and $d_{n+1} = H$. Of course, the levels d_0 and d_{n+1} need possess no critical points, as with Fig. 8.8, which has four critical levels, d_0, d_1, d_2, d_3, but only on d_2 and

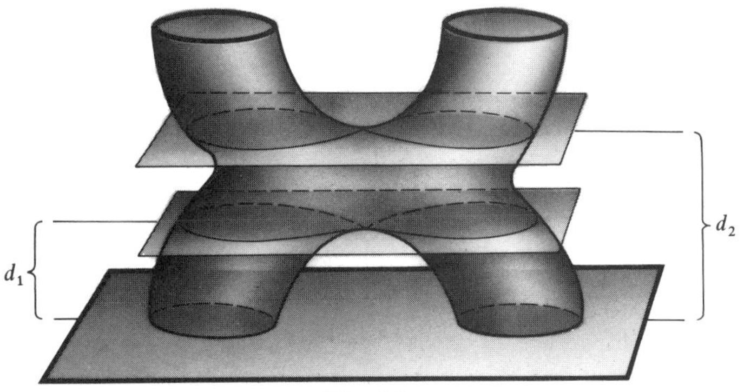

Fig. 8.8

d_1 does there lie a critical point – a pass in this case. Notice that in the figure, the surface T is (in the Family of) a planar region with *three* holes, so $\chi(T) = 1 - 3 = -2$, which agrees with the count

$$\text{pits} - \text{passes} + \text{peaks} = 0 - 2 + 0 = -2,$$

as predicted by the mountaineer's equation.

We have now seen just what the mountaineer's equation means. Notice that it tells us that if $\chi(S)$ is not zero, then there must be some critical points on S (otherwise the left-hand side of the mountaineer's equation would be zero). This shows that critical points are not just accidental; we cannot necessarily get rid of one by reshaping the mountain scenery, without creating another one somewhere else.

We shall prove the mountaineer's equation in the next section.

Exercise 8.2

1. Check the validity of the mountaineer's equation for the surfaces shown in Figs. 8.2, 8.4, 8.5 and 8.8. (For each surface, compute χ and also the number 'pits – passes + peaks'.)

2. Show that if a regular polyhedron (see Section 1.3) is placed on a table, it is not properly arranged unless it rests only upon one corner. Check that the mountaineer's equation is then satisfied.

3. If S is the closed sphere with two handles, use the mountaineer's equation to show that it must have at least *six* critical points, for any proper arrangement of S.

4. If a properly arranged surface S is filled with water to a depth d, draw sketches to illustrate the difference between the surface S^d and the surface of the water. (N.B. The latter is a planar surface.)

8.3 PROOF OF THE MOUNTAINEER'S EQUATION

To prove the mountaineer's equation, let S be a properly arranged surface, and suppose that its critical levels are in increasing order above the horizontal at heights $d_0, d_1, d_2, \ldots, d_n, d_{n+1}$, where we take $d_0 = 0$ and $d_{n+1} = H$ (the height of S). The idea of the proof is to look at the changes that occur in the surface S^d, as the height d of the horizontal slicing plane is increased

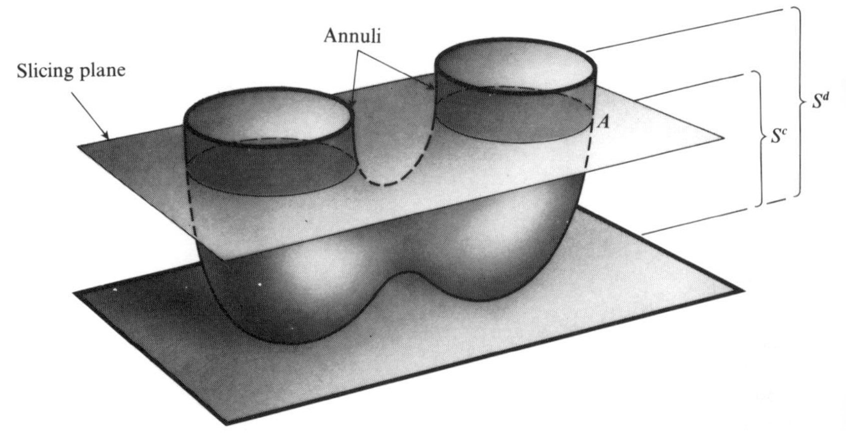

Fig. 8.9. Changes, via ordinary levels are annuli.

from 0 to H. We saw in Fig. 8.4 that S^d may have several components, as did S^a there, but we shall still use the word 'surface' for brevity.

Consider first the case when we have two ordinary levels, c and d, such that $c < d$ and no critical level lies between these two. Then a glance[T] at the contours, as in Fig. 8.9, shows that S^d consists of S^c together with a number of annuli; each such annulus A is joined to a boundary curve of S^c along one boundary of A, while the other boundary of A forms a boundary curve of S^d. But then, as we saw in Exercise 4.7A, No. 2,

$$\text{Family } (S^c + A) = \text{Family } (S^c),$$

so adding *all* the annuli to S^c does not alter the family of S^c. Hence

$$\text{Family } (S^c) = \text{Family } (S^d). \tag{4}$$

Significant changes to S^c, therefore, can only occur when the level c passes through a *critical* level, and we now look at this case.

The zero level is just S^0, and consists of those boundary curves J of S, if any, that lie on the table, together with any absolute minima (pits). If now we raise the slicing plane just a little, to a height c, but below any other critical level, then S^c falls into components – an annulus for each curve J on the table, and a panel for each pit on the table (see Fig. 8.10). In particular then, we have

$$\chi(S^c) = m, \tag{5}$$

the number of separate panels, since $\chi(\text{panel}) = 1$ while $\chi(\text{annulus}) = 0$. We now raise the slicing plane to a height d. As we saw in the previous

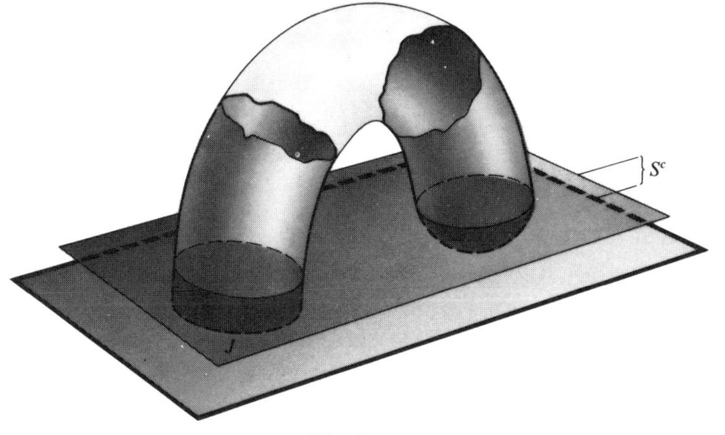

Fig. 8.10

paragraph, the family of S^c does not change as we raise the plane, provided we do not pass through a critical level. Eventually, however, the slicer reaches the new critical level d_1, and we look at S^b, where b is slightly higher than d_1, but lower than the next critical level d_2.

Between the boundary curves of S^c and S^d lie the contours that form the critical level d_1. Upon this level there will lie critical points. If one of these, say X, was a pit, then S^b would contain a panel with centre X, as in

Fig. 8.7; we may assume b so close to d_1 that such panels are separate components of S^b. If the level d_1 contained a saddle point Y, then near Y the picture would look as in Fig. 8.11; there we have drawn in a panel P

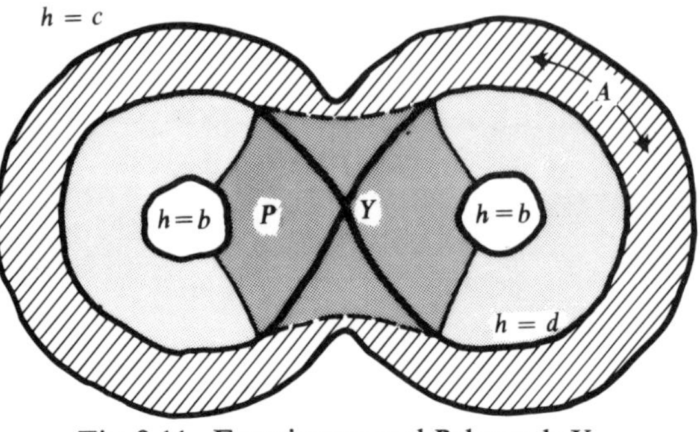

Fig. 8.11. Forming a panel P through Y.

(dark), so that when all such panels are removed from $S^b - S^c$, only annuli A, B, . . . are left. In that figure,

$$S^b - P = S^c + A$$

so

$$\text{Family } (S^b - P) = \text{Family } (S^c)$$

as we agreed before. Now P is added to $S^b - P$ here as an untwisted ear, so in this case

$$\text{Family } (S^b) = \text{Family } (S^c + \text{ear}).$$

A panel can be constructed similarly at every other saddle point on the level d_1, but it might be a bridge between two separate components of S^c (as at L and T in Fig. 8.5).

The critical level d_1 is not high enough to contain a peak, since S is connected (as we shall see shortly), so we conclude that as far as families of surfaces are concerned, we obtain a member of Family (S^b) by taking S^c, adding in some ears or bridges (one for each saddle point), and then adding some separate panels (one for each pit). Therefore we can compute Euler numbers, using Exercise 2.6; so we get

$$\chi(S^b) = \chi(S^c) - s + m,$$

where s is the number of saddle points and m the number of pits on the level d_1. (Addition of an ear subtracts $(-2 + 1)$ from χ, while addition of a separate new panel adds χ(panel), which is 1.)

If we now raise our slicing plane to an ordinary level f below the next critical level d_2, then by the argument for equation (4) above,

$$\text{Family } (S^f) = \text{Family } (S^b).$$

However, as the slicer passes through the critical level d_2, and stops at an ordinary level g just above the level d_2, the same argument as for S^b applies.

That is to say, S^g is obtained from S^f by first adding annuli along boundary curves of S^f. Next, one ear or bridge (possibly twisted) is added, for each saddle point on the level d_1, while for each pit we obtain a new panel as a separate component of S^f. There is one additional possibility, that a hole in S^f is filled up when S^g acquires a panel whose centre is a peak; the nearby contours totally surround it, as at X in Fig. 8.7.

Computing the Euler number, we therefore have

$$\chi(S^g) = \chi(S^f) - s + m + M, \tag{6}$$

where s, m, M are the numbers respectively of passes, pits and peaks on the level d_2. We can now repeat the argument as we raise the slices successively through each critical level; to express the result we need a systematic notation. Recall that the critical levels run from $d_0 = 0$ through to $d_{n+1} = H$, the height of S above the table.

Let then l_j denote the ordinary level $\frac{1}{2}(d_j + d_{j+1})$ which is midway between the jth critical level and the $(j+1)$th, except that l_{n+1} is the highest level H of the surface S. Let L_j denote the surface consisting of all points X on or below the level l_j. Then for each $j = 1, 2, \ldots, n+1$, we have by the same argument as for equation (6) above,

$$\chi(L_j) = \chi(L_{j-1}) - s_j + m_j + M_j, \tag{7}$$

where s_j, m_j, M_j are respectively the numbers of passes, pits and peaks on the critical level d_j.

Equation (7) summarizes the geometrical fact that to obtain the family of L_j we take L_{j-1}, fill in M_j holes and add to it s_j bridges or ears, and m_j new components. (Thus $s_{n+1} = m_{n+1} = 0$.)

Now add all the equations (7) as j runs from 1 up to $n+1$. Then

$$\chi(L_1) + \chi(L_2) + \cdots + \chi(L_{n+1}) = \chi(L_0) + \chi(L_1) + \cdots + \chi(L_n) - s + m + M,$$

where

$$s = s_1 + s_2 + \cdots + s_{n+1}, \quad m = m_1 + m_2 + \cdots + m_{n+1},$$

and

$$M = M_1 + M_2 + \cdots + M_{n+1}.$$

Cancelling out the various χ's, we get

$$\chi(L_{n+1}) - \chi(L_0) = M - s + m.$$

But L_{n+1} is the whole of S. Also, by equation (5) above,

$$\chi(L_0) = m_0 - s_0 + M_0,$$

since $s_0 = M_0 = 0$. Therefore, since $s + s_0$, $m + m_0$, $M + M_0$ are the numbers of passes, pits and peaks on S, we obtain finally

$$\begin{aligned}\chi(S) = \chi(L_{n+1}) &= \chi(L_0) + m - s + M \\ &= (m_0 + m) - (s_0 + s) + (M_0 + M) \\ &= \text{pits} - \text{passes} + \text{peaks},\end{aligned}$$

and the mountaineer's equation is proved.

Note that we have not needed the Assembly Theorem in this chapter, but that we did need the Invariance Theorem to work out $\chi(S)$. Notice too that the slicing process gives us a panelling and assembly for S. If S is closed and *non*-orientable, we cannot physically 'put it on a table' because such *closed* surfaces cannot be constructed in our 3-dimensional world. But the only role of the table was to get a 'height' function easily, and other functions would do, which can be constructed for all surfaces. See, for example, Milnor [13].

Exercise 8.3

1. A man is lost in mountainous country (without ridges or plateaus!) and walks for miles, avoiding pits, peaks and passes, until he returns to his starting point. Show that his route encloses b pits, c passes and d peaks, where $b-c+d=1$ provided he keeps at the same height.

2. Discuss different methods for obtaining a height function on a paper surface. Remember that the table can be tilted, and compare also with 'height above sea-level' on Earth.

3. Show that although $\chi(\text{torus})=0$, a (properly arranged) torus must contain at least two passes.

4. Discuss a possible topography on a torus-shaped planet (or ring of Saturn).

***5.** Show that every closed orientable paper surface S has a height function with just one pit, one peak and $2-\chi(S)$ passes. (Make two holes, and then use induction on an order of assembly to construct a height function which is zero on one hole, 1 on the other, and takes intermediate values everywhere else.)

***6.** Let P be a flat sheet of tin with $n+1$ boundaries, so that it is in Family ($\mathscr{P}_n$). Suppose we charge it up electrically, by putting a high voltage at one electrode on the sheet, and earthing each boundary curve. The graph of the potential can be thought of as a mountainous region built over P, and descending to 'sea-level' (voltage zero) along the boundaries. If the resulting mountain is properly arranged, show that it has just $n-1$ passes ($n \geqslant 1$). [Electrical potentials have neither pits nor peaks inside P, except at the electrode. Also $\chi(P)$ is the Euler number of both P and the mountain.]

9. Miscellaneous exercises

The following exercises are taken from tutorial sheets and examination papers for the second-year Geometry course at Southampton, 1976–9. The wording has been slightly modified, and we include here certain additional instructions (in italics) that were given in the course.

1. One boundary curve of a paper surface A is taped to a boundary curve of a second surface B to form a paper surface C. Show that the number $\beta(C)$ of boundary curves of C is given by $\beta(C) = \beta(A) + \beta(B) - 2$, and that

$$\chi(C) = \chi(A) + \chi(B).$$

(Thus C is a closed surface if $\beta(A) = \beta(B) = 1$, and only in that case.)

2. If $A \sim \mathscr{S}_{p,q,r}$ and $B \sim \mathscr{S}_{u,v,w}$ in No. 1, find the family of C in terms of p, q, r, u, v, w, when C is not closed.

Let S be a paper surface, and suppose we make a small hole in a panel P of S, by removing a small polygonal disc that does not touch the edges of P. Now tape the boundary of the hole to that of a punctured torus T, to form a surface V; join the corners of P to those of the hole by new edges to ensure that V is panelled. We say that V has been formed from S by 'adding a handle'.

3. Find the family of V if $S \sim \mathscr{C}_{q,r}$ (see p. 54).

4. Show that the closed surface $\mathscr{C}_{1,r}$ can be obtained by adding r handles to a real projective plane, and that $\mathscr{C}_{2,r}$ can be obtained by adding r handles to a Klein bottle.

5. A closed surface S is constructed by punching 3 circular holes in a sphere and taping a Moebius band to the sphere along the boundary of each hole. Find the family of S.

6. Show that any closed non-orientable surface $\mathscr{C}_{q,k}$ can be constructed as in the last problem, using $q + 2k$ holes and Moebius bands.

7. (*a*) Show that $\chi(\mathscr{C}_{1r}) = 1 - 2r$ and that $\chi(\mathscr{C}_{2r}) = -2r$.

(*b*) Show that each of the surfaces $\mathscr{C}_{1r}$ and $\mathscr{C}_{2r}$, $r = 0, 1, 2, \ldots$, can be constructed from a sphere by cutting a suitable number of small discs from the sphere and taping the boundary of a Moebius band along the boundary of each hole.

Identify the surface which is formed when two Moebius bands are added to a sphere in this way.

8. Let S_1 be a real projective plane (Moebius band with lid) and S_2 a punctured torus. Remove a small disc D_i from the interior of one panel of S_i (cut a hole in the panel) so that each new surface T_i has a new boundary component J_i ($i = 1, 2$). Add new edges, as for the instructions for adding a handle, to ensure that T_i is a paper surface.

(*a*) Calculate the Euler numbers $\chi(T_1)$ and $\chi(T_2)$.

(*b*) A new surface S can be constructed from T_1 and T_2 by taping J_1 to J_2, to join the surfaces together along the new boundaries. Calculate $\chi(S)$ and identify the Gallic plan of S.

9. Let S_1 be $\mathscr{S}_{0,2,0}$ and S_2 be a torus. Remove a small disc D_i from the interior of one panel of S_i (cut a hole in the panel) so that each new surface T_i has a new boundary component J_i ($i = 1, 2$). Calculate $\chi(T_i)$ ($i = 1, 2$).

A surface S is constructed by identifying J_1 with J_2 (tape T_1 to T_2 along the new boundaries). Calculate $\chi(S)$. Identify and sketch the Gallic plan of S.

10. Let $\mathscr{P}_4$ be the planar region displayed in Fig. 4.9(a). The boundaries of the extreme left and right square holes are joined by a twisted bridge, to form a paper surface S. Calculate $\beta(S)$ and $\chi(S)$ and determine the family of S: (a) by direct application of the Addition Theorem, (b) by displaying a prescription, as in the *proof* of the Theorem.

11. In Exercise 5.3, No. 3 (p. 60), investigate what happens when the four twisted panels are supplemented by k more, all twisted, with opposite ends joined to the boundaries of the octagonal panels. (The latter should be regarded as panels with $8 + 2k$ sides, corresponding edges being joined by the twisted panels.)

12. In No. 11, denote the paper surface by S_k if k twisted panels are used altogether. Suppose one boundary curve of S_5 is joined to one of S_4. Identify the resulting paper surface. What if S_4, S_5 are replaced by S_p, S_q?

13. (a) Show that $\mathscr{S}_{p,q,r} \mathbin{\mathring{+}} P \sim \mathscr{S}_{p-1,q,r}$ when $p \geqslant 1$.

(b) A surface S is constructed by taping the boundary of a Moebius band M along a boundary component of a standard plan $\mathscr{S}_{p,2,r}$ with $p \geqslant 1$ (i.e. $S = \mathscr{S}_{p,2,r} \mathbin{\mathring{+}} M$). Identify the family to which S belongs.

14. Show that the free edges of a paper surface form a set of Jordan curves, no two intersecting.

15. A 'barycentric' subdivision of a panel P is obtained by taking a new corner A inside the panel, and a new one in the middle of each edge. New edges are formed by joining A to each corner, old and new. Show that the Euler number $\chi(P)$ is unaltered by the subdivision.

16. Let S be a panelled surface. A second panelling of S can be obtained by taking a barycentric subdivision of its panels. Show that $\chi(S)$ has the same value for each of these panellings. (If S had n panels calculate how its numbers C, E, P alter under the subdivision, and use the new numbers to calculate χ.)

17. Let J be one component of the boundary of a mystery surface S. Show that by choosing a suitable subdivision of the panels of S it is always possible to construct a neighbourhood N of J on S which is an annulus.

18. Let h be a height function on the properly arranged paper surface S (see p. 85). Define *critical point* and *critical level* for h. State the mountaineer's equation relating the critical points of h to the Euler number of S. Sketch a properly arranged sphere with at least one saddle point, indicating the position and nature of each critical point.

19. S is a closed sphere with two handles. Show that its height function must have at least six critical points for any proper arrangement of S.

***20.** Show that the set of points (x, y, z) in $\mathbb{R}^3$ satisfying the equation

$$4xy = z[4 - (x^2 + y^2 + z^2)]$$

forms a punctured torus, with the puncture 'at infinity'.

The next problems are based on the idea of an 'incidence table' for a paper surface, which is simply a method for describing a surface without supplying a sketch. The idea is most easily seen through a specific example, that of the Moebius band, cut down its middle as in Fig. 1.10. The result of the cutting is a paper surface (an annulus) with two panels P_1, P_2, consisting of the rectangles $adv''u''$, $bcv'u'$ whose edges can be listed

$$e_1 = ad,\ e_2 = dv'',\ e_3 = v''u'',\ e_4 = u''a \quad (\text{on } P_1),$$
$$f_1 = bc,\ f_2 = cv',\ f_3 = v'u',\ f_4 = u'b \quad (\text{on } P_2).$$

Thus, e_2 is taped to f_4, and e_4 to f_2, leaving us with six distinct edges e_1, e_2, e_3, e_4, f_1, f_3 and four distinct corners a, d, u', v' (since $b = d$ after the taping). Now draw up the following incidence table:

	P_1	P_2	a	d	u'	v'
e_1	1	0	1	1	0	0
e_2	1	1	0	1	1	0
e_3	1	0	0	0	1	1
e_4	1	1	1	0	0	1
f_1	0	1	1	1	0	0
f_3	0	1	0	0	0	1

The edges are listed along the side, and the panels and corners along the top. In the row labelled by an edge e, we insert a '1' under each panel (or corner) that e lies in (or lies in e) respectively; otherwise we write '0'. For example, the table here tells us that e_1 lies in P_1 but not in P_2, but e_4 lies in both panels, while e_1 has ends a, d. (Note: the labelling need take no account of any orienting arrows.) Given an incidence table, the surface can be reconstructed by starting at a corner, drawing the edges that issue from it, moving to the other ends of these to find which edges issue from them, and so on. It is easy to calculate the Euler number, simply because the table displays the numbers of corners, edges and panels; and since a free edge has only one '1' beneath a panel we can pick out all free edges and determine the number β of boundary curves. We can then determine a plan $\mathscr{S}_{p,q,r}$ using the procedure of p. 61.

21. Suppose that in Fig. 1.10 the arrow on cv' is reversed. Draw up the resulting incidence table (which will describe a Moebius band).

22. Draw up the incidence tables for a triangular panel, and for a cube.

23. Construct a paper surface S with incidence table as follows; the v's are corners.

	P_1	P_2	P_3	P_4	v_1	v_2	v_3	v_4
e_1	0	0	1	1	0	0	1	1
e_2	0	1	0	1	0	1	0	1
e_3	0	1	1	0	1	0	0	1
e_4	1	0	0	1	0	1	1	0
e_5	1	1	0	0	1	1	0	0
e_6	1	0	1	0	1	0	1	0

24. Identify the surface whose panelling has incidence table:

	P_1	P_2	P_3	v_1	v_2	v_3	v_4
e_1	1	0	1	0	1	0	1
e_2	0	1	0	0	1	1	0
e_3	1	0	1	1	0	1	0
e_4	0	1	0	1	0	0	1
e_5	1	1	0	0	0	1	1
e_6	1	1	0	1	1	0	0
e_7	0	0	1	0	0	1	1
e_8	0	0	1	1	1	0	0

and draw its Gallic plan (p. 61).

25. Calculate the Euler number and the number of boundary components for the surface S whose panelling has incidence table:

	P_1	P_2	P_3	P_4	v_1	v_2	v_3	v_4
e_1	1	1	0	0	0	0	1	1
e_2	0	0	1	1	0	0	1	1
e_3	1	1	0	0	1	1	0	0
e_4	0	0	1	1	1	1	0	0
e_5	1	0	1	0	0	1	0	1
e_6	0	1	0	1	0	1	0	1
e_7	1	0	1	0	1	0	1	0
e_8	0	1	0	1	1	0	1	0

and hence identify the surface.

Appendix A. Mathematical theory of surfaces

The purpose of this appendix is to outline a rigorous mathematical foundation for the theory expounded in the text. It is assumed here that the reader is acquainted with the elements of topology (such as in Griffiths and Hilton [7], ch. 25) and can read some of the mathematical texts which we cite. To save space, proofs of results are omitted, and instead we give sources where they may be found.

A.1 NOTATION

We use $\mathbb{R}$, $\mathbb{N}$ respectively to denote the sets of real, and natural, numbers. In the Euclidean plane $\mathbb{R}^2$, we let S^1 and D denote respectively the unit circle $x^2 + y^2 = 1$ and the unit disc $x^2 + y^2 \leqslant 1$. Each is compact. If $n \in \mathbb{N}$ and $n \geqslant 3$, we let D_n denote the unit disc together with its *vertices*, the n points with angular coordinates $2\pi k/n (k = 0, 1, \ldots, n-1)$. A *Jordan curve* in a metric space X is a mapping $f: S^1 \to X$ that is a homeomorphism onto its image $f(X)$. (It suffices to say that f is a continuous injection, since $f(X)$ is compact.) We do not distinguish here between f and $f(X)$.

A.2 JORDAN AND SCHOENFLIES

Any rigorous theory requires statements and proofs of the Jordan Curve Theorem ('Every Jordan curve separates the plane $\mathbb{R}^2$ into exactly two components, of which it is the common frontier') and the Schoenflies Theorem ('Given a Jordan curve in $\mathbb{R}^2$, there is a homeomorphism $f: \mathbb{R}^2 \to \mathbb{R}^2$ which carries the curve onto the unit circle, and hence its interior onto the unit disc'). Proofs are in Newman [14], pp. 137 and 173. For an illuminating elementary proof of the polygonal Jordan Curve Theorem, see Courant and Robbins [5], ch. 5. Note that the Moebius band cannot be planar, because of the Jordan Curve Theorem; for, its centre-line (a Jordan curve) does not separate it (see Section 1.5).

A.3 MATHEMATICAL SURFACES

Mathematical models[T] of the panels and paper surfaces of the text can now be made as follows. A 'paper panel in our 3-dimensional world' can be modelled by a continuous injection $p: D_n \to \mathbb{R}^3$ (for some $n \in \mathbb{N}$, $n \geqslant 3$). Reference to $\mathbb{R}^3$ can be eliminated, if desired, by taking 'a paper panel P' to correspond to a homeomorphism $p: D_n \to P$, where P denotes a topological space. The *edges* of the panel then correspond to the restrictions $p|e_i$, where e_i is the ith edge of D_n. The *vertices* are the images of those of

D_n. More generally, a 'lamina' is again a (mathematical) panel. (We shall use the adjectives 'paper' and 'mathematical' to distinguish the objects in the text from their mathematical models.)

Corresponding to 'paper surface in $\mathbb{R}^3$', we now have a mathematical model based on the insight of Exercise 7.1, No. 5. Thus we consider a connected subspace $S \subseteq \mathbb{R}^3$ together with mathematical panels $p_i: D_{n(i)} \rightarrow S$, $i = 1, \ldots, s$, such that S is the union of the images $P_i = p_i(D_{n(i)})$ and where (i) two such panels intersect only along common edges or vertices, (ii) no edge is an edge of more than two panels, (iii) the Star of each vertex is homeomorphic to some D_n. A short proof establishes easily that the boundary of a mathematical surface consists of a finite family of mutually disjoint Jordan curves, composed of all the free edges of the panelling. Therefore, this definition corresponds to that of 'mystery surface' in Chapter 7, but the Assembly Theorem proved there applies directly here (with obvious re-coding from 'paper' to mathematics) to yield an order of assembly for S. To triangulate such a surface, we divide D_n by n radii from the centre O to its vertices; the images of the resulting triangular discs are the triangular panels of S.

If reference to $\mathbb{R}^3$ is required to be eliminated, we can define a surface more abstractly as the quotient space of a disjoint union of panels, modulo 'glueing' homeomorphisms g_{ij}^{ab} from the ith edge of panel P_a to the jth edge of P_b. This models the 'taping' process. Conditions (ii) and (iii) above must also be imposed.

Of course, the instructor may want to go the whole* hog and take, as model of a paper surface, a compact connected Hausdorff space in which every point has a neighbourhood homeomorphic either to $\mathbb{R}^2$ or to the upper half-plane $y \geqslant 0$. The points of the second kind form the boundary of S. One must then show how to triangulate S, using (say) the proof given in Ahlfors and Sario [2], Chapter 1, Section 45C. This is technical, and uses the Schoenflies Theorem. It will not be understood by many pupils who may well be mature enough for the 'paper' theory.

By such various routes, however, we obtain a (triangulated) mathematical model of the paper surface.

A.4 FAMILIES

If S, T are two (mathematical) surfaces, we write Family (S) = Family (T), if, and only if, S and T are homeomorphic. It then becomes necessary to establish the 'Agreements' of the text as theorems about (triangulated) mathematical surfaces.

It is an instructive exercise for beginners, to start by proving Agreements 2, 3 and 4 for mathematical panels. These concern fairly easy extensions of homeomorphisms, which beginners nevertheless find hard. Much more difficult is a proof of Agreement 7, which depends on the Schoenflies Theorem; see Ahlfors and Sario [2], Chapter 1, Section 45B.

* Or even further, by adopting the definition of 'generalized manifold' and proving the locally Euclidean property in dimension 2: see Wilder [19], p. 272, th. 2.3.

Next, consider Agreement 1, that $S \dotplus D$ is homeomorphic to S when S is a (mathematical) surface and D a panel such that $D \cap S$ is an arc α (divided into certain edges of D). By constructing an arc β in S which meets the boundary of S only in the end-points of α, and such that $\alpha \cup \beta$

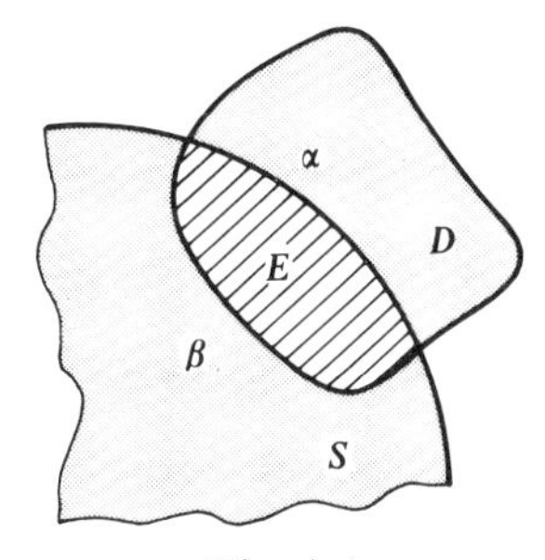

Fig. A.1

encloses a topological disc E, we then construct a homeomorphism $g: E \cup D \to E$, which is the identity on β. Hence g extends as a homeomorphism $S \cup D \to S$ as required.

Using such extensions of homeomorphisms of discs, we can establish Agreements 5 and 6 for mathematical surfaces. (See also Exercise 4.4**B**, No. 4.)

A.5 MATHEMATICAL LANGUAGE

Now, the deductions made in the text are based on the seven Agreements. We have just seen that these Agreements can be established for mathematical surfaces and hence all the deductions made in the text are valid for them also. One might feel that arguments like that of the Addition Theorem (Section 4.5), which involves drawing many figures, should be given in a purely symbolic way, without pictures; then they would look less naïve, more 'mathematical'. In fact, the figures *could* be described using only words and mathematical symbols, but the resulting description would be unreadable. Such arguments should be written and practised privately by enthusiasts; mathematicians rarely publish every detail of a long proof. (See, for example, Wang [18] who tells just how difficult it is to put even quite elementary arguments in fully detailed, computer-checkable form – the ultimate test of a detailed argument.)

A.6 FURTHER OUTLOOK

The Classification Theorems for surfaces, while interesting in themselves, and outstanding examples of the classification theorems of mathematics, were never designed as ends in themselves. Historically, it was necessary to formulate, and then prove them, in order to simplify problems of analysis (calculus) and differential geometry. We have been considering the *topological* structure of compact surfaces, but mathematics requires us to show that every such surface has a differentiable (indeed real-analytic) structure and that every orientable surface has a complex-analytic structure (see Ahlfors and Sario [2], p. 127). In the latter case one can investigate the Riemann theory of functions of a complex variable on a surface (see, e.g. Klein [12]) and Poincaré's theory of automorphic functions. For that purpose, Poincaré introduced the 'fundamental group' of a surface, and the theory of covering surfaces (which need not be compact). This turns out to be the theory of plane non-Euclidean (hyperbolic) geometry, with the associated Teichmueller theory (see Ahlfors et al. [1]).

From the 'differentiable' point of view, there is the differential geometry of surfaces, which considers such items as their curvature properties (see Willmore [20]), and the smoothing of homeomorphisms. There is also the theory of vector fields and differential equations on a surface (see Arnold and Avez [3]). Surfaces are indeed a rich field for investigation, and the results already found are the starting points for studying higher-dimensional objects, called 'manifolds'.

For example, a 3-dimensional manifold can be defined combinatorially, analogously to a surface, but using tetrahedra instead of triangles; but now every triangle is to be a face of at least one tetrahedron, and of not more than two. Examples are the familiar solids whose surfaces are closed spheres with g handles, and the exterior of a knotted rope in a solid cube. No classification theorem, analogous to that for surfaces, has yet been discovered for 3-manifolds – in spite of a great effort by many mathematicians and the invention of clever and powerful techniques. Apart from research papers, the best exposition still seems to be that in the book of Seifert and Threlfall [17].

An n-dimensional manifold can be defined as a Hausdorff space in which each point has a neighbourhood homeomorphic either to $\mathbb{R}^n$ or the half-space $x_n \geqslant 0$. Surprisingly, it is when $n \geqslant 5$ that much progress has been made with the investigation of n-manifolds. An important tool of investigation has been the Morse Theory, in a refined form (but in n dimensions) of the material of Chapter 8. Little is known about 4-dimensional manifolds, and the methods that work well when $n \geqslant 5$ are not successful when $n = 3$ or 4. But the latter pair are the dimensions needed for models of our own space–time, so they are of greatest importance outside mathematics. And it is perhaps not too much to say that we shall never crack open the problems of 3 and 4 dimensions until we have a really deep understanding of surfaces, and especially of the way in which curves can lie upon them (a topic we have not pursued in this book). For a survey see, e.g. Hilton [10].

Since this book was first printed, a good, readable, introduction to many of the above ideas has appeared – *Basic Topology* by M. A. Armstrong (McGraw-Hill, 1979). For a fascinating discussion of the history of Euler's formula $\chi(\text{cocoon}) = 2$, see *Proofs and Refutations* by I. Lakatos (Cambridge University Press, 1976); this is a 'must' for all interested in Mathematics Education, and even more essential for academic mathematicians.

Appendix B. Teaching notes

These notes are intended for two kinds of reader. First, they are for those who may wish to try to teach children about surfaces, once they themselves have learned the material of the book. Second, they are for those who know 'academic' topology well, but who may wish to teach undergraduates from the text. It is difficult not to sound patronizing to one kind of reader or the other, and the author asks the reader's indulgence.

p. 1. This question is intended to be put directly to the class by the teacher, who should write down on the blackboard any definition offered by a pupil. The resulting discussion may throw light for the pupils, on how and why definitions are formulated by mathematicians (i.e. people). It is the nature of a book that, for brevity, 'the' definition is simply written down. While it is unlikely that an audience will produce any different workable definition (provided those in the know have been asked to keep silent), the teacher should be prepared to respond to such a definition, to see how a theory of it might compare with the theory of the text.

p. 2. We have used the plus sign rather than the set-theoretic 'cup' for simplicity, for pupils unfamiliar with the latter.

p. 7.(*a*) This may be the pupil's first meeting with a formal definition, and it should not be taken for granted that he will take it at face value; a clear definition can seem vague at first. The use of such words as 'hence' and 'therefore' should perhaps be discussed as well, at this stage.

p. 7.(*b*) A course for undergraduates might well be given which begins with graph theory and which leads to the theory of surfaces using as motivation the planarity of graphs. For example, what is the closed surface of least genus in which a given graph can be embedded?

p. 10. It is probably best in class not to raise these points about multiple twists at first, but the presence of left-handed pupils may raise the question of direction of twist, in a natural way. However, our reason for ignoring the direction, and thinking of only one (or no) twist, is that we would otherwise be concerned with the technically advanced theory of classifying all the ways in which a given surface can lie in the surrounding space. For example, in Figs. 1.7(*a*) and 1.7(*d*), we see two quite different ways in which an annulus can lie in its surroundings: the over- and under-passes in Fig. 1.7(*d*) cannot be removed without breaking the annulus. We are therefore ignoring these complexities because they are not relevant to the theory we wish to develop; and any pupil who is uneasy will have to be asked for patience, to do the easier thing first! As far as the problem is concerned, of classifying the ways in which a surface can lie in our surrounding space, nobody yet knows a complete answer; for we would need a complete list of all essentially different knots, and such a list has not yet been obtained.

p. 14. Here and elsewhere, we refer intuitively to the boundary curves, to avoid (at this stage) a cumbersome definition and proof. The point can however be brought out in verbal discussion.

p. 15. This complication, which may well be raised by pupils when they make models, should be treated in the spirit of the note for p. 10. It will apply also to the addition of bridges (see p. 17), and will be resolved when we consider Agreement 5 (see the Remark on p. 41).

p. 16. This is meant to be observed empirically. A mathematical discussion involves the Jordan Curve Theorem (see Appendix A, Section 2), which was first pointed out by the mathematician C. Jordan.

p. 18. The words 'corner' and 'number' are used as being more familiar than 'vertex' and 'characteristic' (so less distracting to a pupil).

p. 21.(*a*) The associativity of + is silently passed over here as being obvious, and a discussion at this stage would be distracting. A pupil can 'know' without having to 'say' or be told.

p. 21.(*b*) The construction cannot be performed in 3-dimensional (Euclidean) space, by Alexander duality in Homology Theory. An intuitive proof in Exercise 3.5, No. 5 depends on a separation property, which needs duality to prove it.

p. 26. This may be a good place to discuss, with pupils, associativity and the need for brackets, and perhaps commutativity of the various plus signs.

p. 28. The dogmatic assertions of the text (made for brevity) should of course be introduced by a democratic discussion.

p. 29. There is of course a profound difference between 'If we can manage it' and 'If it can't be done'. This difference and its implications could form the basis of an interesting discussion.

p. 30. A discussion about the 'differences' between surfaces is essential. It should be in the form 'Would you regard such and such as different?' rather than the authoritarian 'I will regard the following differences as negligible'.

p. 32. It seems less authoritarian to emphasize family resemblances rather than neglect of differences (which may seem highly non-neglectable to beginners). The Agreements which follow are consequences of forming equivalence classes of homeomorphic surfaces (see Appendix A, Section 4); but they are deliberately framed as written, as likely to be more intelligible to beginners who need not then also make the intellectual jump to the concepts and language of 'official' topology. We are therefore setting up the Agreements as 'Pedagogical Axioms' in the sense of Griffiths and Howson [8], Chapter 16, because logical economy is not a primary aim here: understanding by the pupil is what we wish to achieve.

p. 34. 'Cocoon' is used, as being less distracting than '(topological) sphere'. To a beginner, spheres are round.

p. 35. Here and elsewhere, a formal proof requires mathematical induction. When experience and the need for it have developed, the method of induction is formally discussed on p. 59. At the present stage, however, it suffices if the pupil sees the formulae intuitively (essentially by proving 'P_n implies P_{n+1}') without setting up the formal apparatus.

p. 36. This is an example of what Bourbaki calls an '*abus de langage*'. It should not be used without explanation, because the shorthand that is essential for doing mathematics can also be confusing to beginners who are at the same time learning to make fine distinctions.

p. 38. We are here planting the word 'plan' in order to use it formally later. A plan of a house (say) emphasizes the 'important' features without pretending to be an accurate representation. This point about deliberate suppression of 'noisy' detail should be discussed verbally (see Griffiths and Howson [8], p. 3).

p. 39.(*a*) 'Its plan $\mathscr{P}_n$' is an abbreviation (which needs explanation to beginners) for 'It has a plan, which we always denote by $\mathscr{P}_n$'. Pupils may also need a word about the use of different founts of type. This point also occurs later with the plans $\mathscr{T}_n$ and $\mathscr{S}_{p,q,r}$.

p. 39.(*b*) If pupils object to Agreement 5, they should be encouraged to prove (or disprove it) to see what mathematics they generate. (See Exercise 4.4B, No. 4.)

p. 43. In this problem and the next, we have told the pupil the answer and he must find reasons. In class-work it is hoped that the pupil will be asked, by his friends, questions like those in the third problem of this set. Awareness and curiosity must be stimulated.

p. 44. A teacher should expect to spend some time, depending on the maturity of the class, on the use of this symbol.

p. 45. The exposition chosen here is a compromise between two extremes. On the one hand, we could discuss many individual cases, with specific numbers chosen for p, q and r. On the other hand, we could use bijections to bring out a precise notion of isomorphism between the two prescriptions. Pupils of differing maturity require different approaches, to be judged by the teacher, but the material of the paragraph is so crucial to an understanding of what the mathematics of surfaces is about, that a teacher may prefer to use both the extremes and our compromise at different stages. He would then be working in accord with the 'spiral approach', using repetition but rising through stages as the pupils gain more meaning from work with the textual material.

p. 46.(*a*) A change of letter may be less confusing (but incur a cost in memory) to a pupil than the use of primed letters (which he may have met only in calculus as with $f'(x) = df/dx$). However, an important part of his education is to get used to different alphabets.

p. 46.(*b*) When an apparent 'trick' occurs in a piece of mathematics it is important to try to show how it was discovered, otherwise a pupil may be depressed that he could never find one, or repelled by its arbitrary nature: 'If God intended us to discover truths, he surely did not intend us to stub our toes on them'. And of course, in geometry one cannot emphasize too strongly the need for experiments. Mathematics is, among other things, an experimental science.

p. 48. Case-by-case analysis is foreign to most of those students who know only* A-level mathematics. Certain of the exercises depend also on this technique, which is why they look very easy to sophisticated mathematicians and yet may be *paralysing* to beginners.

p. 51. See note to p. 46 about 'tricks'.

p. 52. Note associativity of '+' (see note to p. 26).

p. 53. Strictly we should say 'Let L denote . . .' rather than 'Let L be'. This pedantic point is covered up in such usage as 'Add the panel B'.

p. 54.(*a*) We use $\mathscr{C}_{q,r}$ for the family rather than for a specific element of it, to avoid a complicated description of a *standard* closed surface. Such a description is much more complicated than that of $\mathscr{S}_{p,q,r}$, particularly because we cannot point to a physical model of a closed non-orientable surface, unless we allow distracting singularities as in Fig. 2.8.

p. 54.(*b*) Sketching should be emphasized: it can give a lot of insight into the difference between the geometry of the plane and that of space.

p. 55. A mathematician's instinct would lead him to 'clarify' this description, by using structure-preserving bijections. A pupil, however, may receive more of the meaning if he does not also have to grapple with (for him) extraneous objects like bijections. When he has digested that part of the message that wordy passages have given him, he may well then ask for clarification. He may then see the point of talking about bijections. (Incidentally, he may well have known what a bijection 'is' for a long time: the problem is that he perhaps failed to understand what the definition of a bijection *says* – how it models his concept.)

p. 58. Here again mathematical induction is implicit (see next note).

p. 59. Because of this point, we decide to introduce mathematical induction in a formal way in the next paragraph. Most pupils will still probably find the earlier ('plausible') explanation more convincing, because mathematical induction so often appears to them (when *formally* stated) either (*a*) unintelligible or (*b*) a soufflé that proves nothing. A mathematician's clarity is 'woolly' to unsophisticated minds. These teaching difficulties account for avoidance of the phrase 'E_m implies E_{m+1}' and use of such imprecise phrases as 'If E_m holds' and 'If we could prove E_m' or even 'If E_m is true'. The method really has to be allowed to speak for itself, but requires mature ears for listening.

* Non-British readers may care to know that 'A-level' refers to a public examination taken by mathematics specialists at the end of their high-school careers. For sample examination papers, see Griffiths and Howson [8].

p. 60.(*a*) This may be the first time the student has met an argument based on the notion of parity.

p. 60.(*b*) An answer is merely to find *one* plan, but a mathematical approach (as opposed to an examination approach) is to find *all* possibilities. The necessity for considering other possibilities should need no stressing (in life as well as in mathematics) but few A-level mathematicians have acquired the habit.

p. 60.(*c*) This dogmatic statement saves space in a book, but should not be asserted to a class until they have been asked their considered opinion.

p. 61. An ordered assembly is not needed for the theorem to hold, but only for pedagogical reasons to simplify the proof. Later we show how to dispense with the need (see the Assembly Theorem in Section 7.2). There is surely no need to disbar many people from knowledge of the Classification Theorem because they may lack the stamina needed for the arguments of Section 7.2.

p. 64. Here is a good example of the need in mathematics to generalize, in order to isolate the essential features of a problem. This is analogous to the way in which one sometimes solves practical problems about measuring with real numbers, by using the bigger system of complex numbers; e.g. if a quadratic equation occurs, one finds its roots *knowing that they exist* as complex numbers, and further argument must then be given to decide on their precise properties.

p. 65. Strictly speaking, the Euler numbers should be denoted by $\chi_P(S)$, but we simplify the notation if no ambiguity is imminent. The use here clashes with our normal usage of the symbol χ_S in Section 2.6 and elsewhere. Notice the use of $\chi(C)$ in the discussion lower down.

p. 66. For a broad discussion, see Armitage and Griffiths [4], Chapter 5.

p. 67. This is a strong form of Agreement 4, and a formal proof depends on the Schoenfliess Theorem (see Appendix A).

p. 69. See note to p. 59.

p. 71. It is tempting to use new letters, say script, for the various panellings. However, this would introduce a possibly distracting notation, and pupils may themselves get so tired of reading '-panelling' etc. that they suggest a notation! The entire proof as written is long-winded, but if it is compressed it becomes opaque because of the number of new ideas in it.

p. 72. Some (simple) logical steps are omitted here, to give pupils practice in filling them in.

p. 76.(*a*) Pupils should be asked to supply their own example. This is an instance of the technique of proof by counter-example, which will be new to many pupils.

p. 76.(*b*) The 'official' definition of a combinatorial surface is avoided here because of the difficulty (at this stage) of proving that each boundary component is a Jordan curve. Condition (ii) is here regarded as basic and intuitive. But see Exercise 7.1, No. 5.

p. 78.(*a*) The proof sketched in Exercise 6.6, No. 10, has an intricate logical structure, in spite of its simplicity. Many pupils may be satisfied with the sketch, but some may need a discussion of the Contrapositive in Logic; because the proof uses the form: 'If $g_R < n - 1$, then we can increase g_R; but we can't increase g_R, so $g_R = n - 1$.'

p. 78.(*b*) Pupils should be asked to suggest what theorem might be provable here, before presenting it cut and dried.

p. 79. A flow chart might make the logical procedure clearer.

p. 80. Pupils may here need to be told what 'if, and only if' means in mathematics. They can learn it from experience in the argument lower down, and we have earlier 'planted' it informally, where its precise meaning did not matter so much. We also have used earlier 'when, and only when'.

p. 84. The usual statement of the mountaineer's equation concerns the critical points (in the sense of calculus) of any twice-differentiable function f on the surface S, when S is provided with a differentiable structure. But then the graph of f on S can be embedded in Euclidean 5-space and f becomes the height function. For this reason we have restricted ourselves to the height function in the text, at the cost of losing a bit of generality, but at a gain of considerable simplicity.

p. 87. For the genuine paper surfaces of our experience, this *is* clear, but of course we are silently invoking the (intuitively obvious) theorem of the maximum for continuous functions defined on compact sets (see Griffiths and Hilton [7], p. 427) as well as the Implicit Function Theorem (see Armitage and Griffiths [4], p. 158).

p. 91. Here, we *mean* a glance: work with a contour map could be a good preliminary to the work of this section. A technical proof for a smooth surface uses the orthogonal trajectories of h: see Milnor [13].

p. 95. Ideas of mathematical modelling can be introduced to students along the lines of Hirst and Rhodes [11]. See also Griffiths and Howson [8].

Exercises

1. Devise a conventional examination paper on surfaces (for* CSE, O-level, A-level), as appropriate. Supply solutions.

2. How would you assess a person's understanding of surfaces?

* These abbreviations refer to British public examinations. Non-British readers should substitute their own public examinations instead.

Appendix C. Hints and solutions to exercises

Exercise 1.3 (p. 7)

In Fig. 1.2(a) and (b) the number $C-E+P$ is 1, in Fig. 1.2(c) it is 2 for each surface, while in Fig. 1.3 it is 0 for the two toruses, –6 for the surface of genus 4, –3 for the double torus with hole, –4 for the block with 3 tunnels, 1 for the open box and 0 for the remaining surfaces.

Exercise 1.6 (p. 12)

***2.** The sets are sketched in Fig. C.1 and consist of (a) a pair of parallel planes, (b) a circular cylinder, (c) a sphere, and (d) a double cone, a hyperboloid and a 'hyperboloid of 2 sheets'. Only the sphere can be modelled,

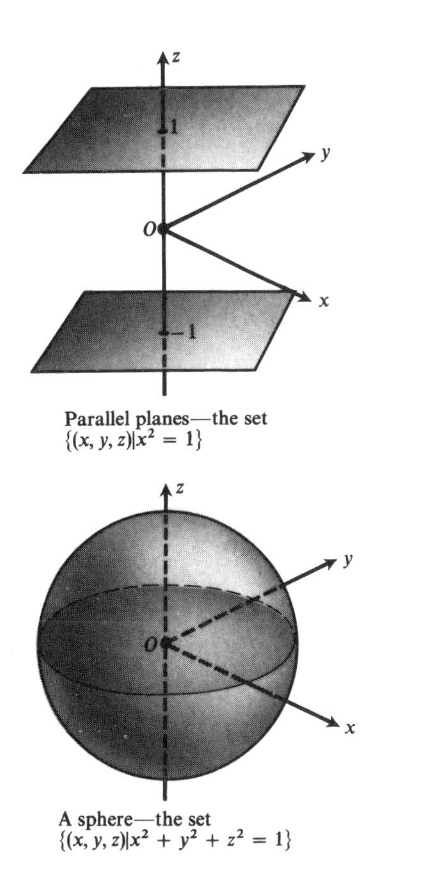

Parallel planes—the set $\{(x, y, z)|x^2 = 1\}$

A sphere—the set $\{(x, y, z)|x^2 + y^2 + z^2 = 1\}$

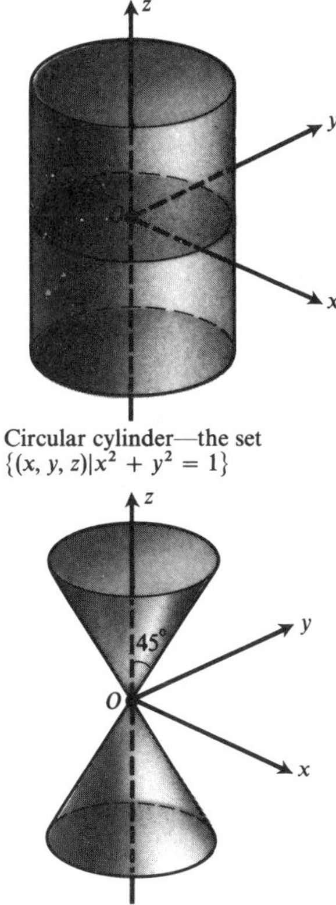

Circular cylinder—the set $\{(x, y, z)|x^2 + y^2 = 1\}$

A double cone—the set $\{(x, y, z)|x^2 + y^2 - z^2 = 0\}$

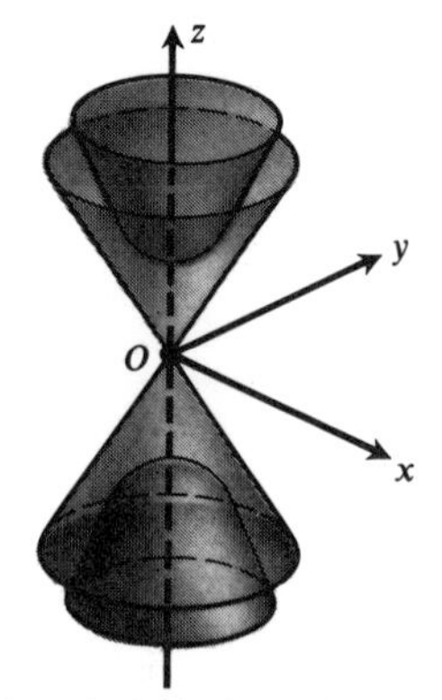

The hyperboloid of two sheets —the set $\{(x, y, z) | x^2 + y^2 - z^2 = -1\}$

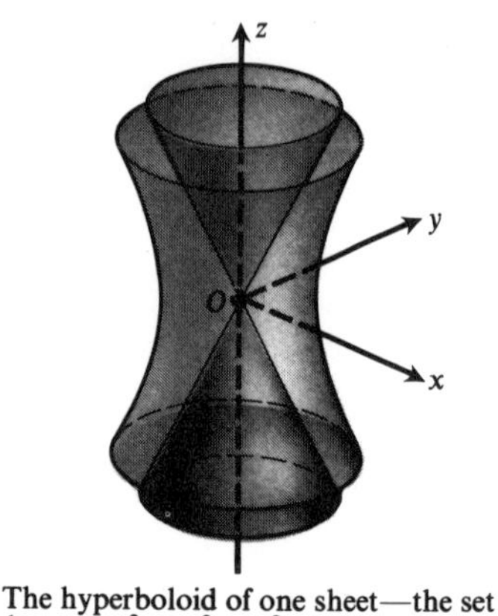

The hyperboloid of one sheet—the set $\{(x, y, z) | x^2 + y^2 - z^2 = 1\}$

(The cone is for comparison, not a part of either hyperboloid.)

Fig. C.1

by paper surfaces if we insist on our rules, since the other surfaces sketched are of infinite extent. (Technically speaking, only the sphere is 'compact'.)

***3.** Suppose $P=(x,y,z)$ lies on the torus. Let the vertical plane parallel to the z-axis, through the centre of the torus and through P, cut the torus in the circle shown. Then in the notation of Fig. C.2 – where a is the radius

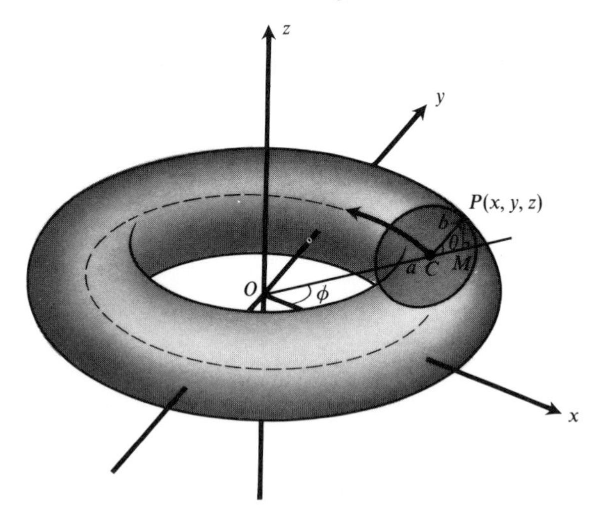

Fig. C.2

of the central 'core' and b is the radius of the circular section – $x=(a+b\cos\theta)\cos\varphi$, $y=(a+b\cos\theta)\sin\varphi$. But $\sin\theta=PM/CP=z/b$, whence $\cos\theta=\sqrt{(b^2-z^2)}/b$. Since $\sin^2\varphi+\cos^2\varphi=1$, then $x^2+y^2=(a+b\cos\theta)^2$ and substituting for $\cos\theta$ we have an equation of the form $f(x,y,z)=0$. (It is not very manageable for most purposes.)

★4. (*a*) represents a hyperboloid (as in 2(*d*) above) which cuts the plane $z = 0$ in the circle $x^2 + y^2 = a^2$; indeed it cuts any plane $z = k$ in the (larger) circle $x^2 + y^2 = a^2 + k^2$. (*b*) represents a sphere, centre the origin, radius b. (*c*) Since $z^2 \geqslant 0$, any point (x, y, z) on the surface must be such that $1 \geqslant r^2 \geqslant c^2$. Moreover, if $z = 0$, then this point is either on the circle $r^2 = 1$ or on $r^2 = c^2$ since either $1 - r^2 = 0$ or $r^2 - c^2 = 0$. Also for any $P = (x, y, z)$ on the surface, $z^2 \leqslant (1 - r^2)(1 - c^2)$ so P lies inside (or on) the unit sphere $r^2 + z^2 = 1$. The surface is symmetrical about the (x, y)-plane $(z = 0)$ and hence is a torus whose 'hole in the middle' surrounds the circle $x^2 + y^2 = c^2$.

★5. Arguments similar to those for (*c*) in the last solution show that the equation represents a double torus as in Fig. C.3. To obtain more holes, we include more factors in the product for z^2, of the form$[(x - c)^2 + y^2 - w^2]$, or with y replaced by $y - d$ if we wish to have holes off the x-axis.

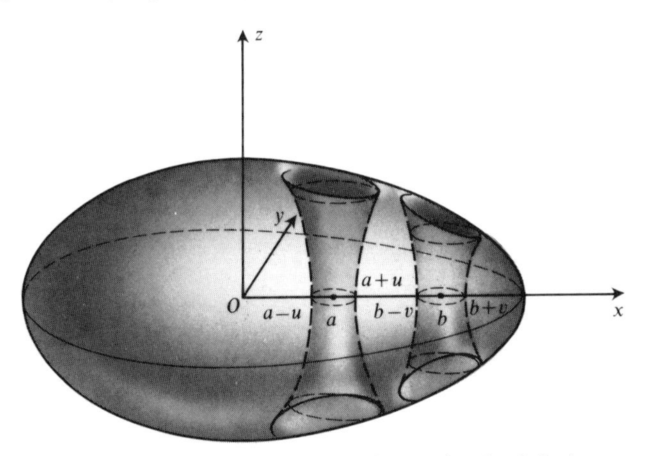

Fig. C.3. A double torus; approximate sketch of the locus

$$z^2 = (1 - r^2)[(x - a)^2 + y^2 - u^2][(x - b)^2 + y^2 - v^2].$$

★6. For a full description, see p. 379 of Vol. 2 of *Differential and Integral Calculus*, by R. Courant (Blackie, 1952).

Exercise 2.2 (p. 15)

1. 6; **2.** $1 + h$.

Exercise 2.3 (p. 16)

1. None (trace round the curve in Fig. 2.3).

2. Only the untwisted ears affect the number of boundary curves.

Exercise 2.5 (p. 18)

3. Ends of a bridge must lie on different boundary curves.

4. A panel with 8 ears has 9 boundary curves, so (since addition of a bridge reduces the number of boundaries by 1) we can add at most 8 bridges. This would form a surface with just 1 boundary curve, to which no more bridges may be added.

Exercise 2.6 (p. 19)

2. 4 ears would do; but for a systematic treatment, see No. 5 below.

3. When a bridge or ear, with n edges, is taped to a surface, then $n-4$ new corners and $n-2$ new edges are added, while the number of panels increases by 1. Thus χ changes by $(n-4)-(n-2)+1=-1$.

***5.** Using Nos. 3 and 4, if u ears and v bridges are added to a panel, then the resulting surface has $\chi = 1-u-v$. Thus if $\chi=-3$, then $u+v=4$. The possible pairs (u,v) are then $(u,v)=(0,4)$, $(1,3)$, $(2,2)$, $(3,1)$, $(4,0)$. Corresponding to each pair, we can decide how many twisted panels to allow; for example, with (1, 3) we could have 1 ear twisted and no bridges, or 1, 1, or 1,2, or 1,3 – or with no ear twisted and no bridge, or 0 and 1, or 0, 2, or 0, 3. The other pairs (u, v) are treated similarly.

6. $\chi_S = 1-21-13=-33$ (by No. 3 above). If t ears are untwisted, and since we need at least two boundary curves in order to add a bridge, then $\beta_S = 1+t-13 \geqslant 1$, so $t \geqslant 13$. Hence β_S is at most $22-13=9$ and at least 1, depending on t.

***7.** The '-2' and '-1' in the formulae for C_V and E_V arise because we would be counting elements of the common edge twice if we simply formed C_S+C_T and E_S+E_T. The formulae for β_V and χ_V are obtained by using the expressions for C_V, E_V, P_V and regrouping.

8. Using No. 3 above, when u ears and v bridges are added to a panel, $\chi = 1-(u+v)$. If we then add a lid, we leave the numbers of corners and edges unchanged, but the number of panels increases by 1, to make $\chi = 2-u-v$, and now the surface is closed.

***9.** If v_i, v_j are the ends of one edge e, then e is counted twice in the sum $M(v_i)+M(v_j)$ (and in no other term $M(v_k)$). All edges are thus counted twice, to give $2E$. Suppose the v's are numbered so that $v_1, v_2, \ldots, v_r$ are the corners with $M(v)$ *even*. Then $m = M(v_1)+\cdots+M(v_r)$ is an even number, and hence so is $2E-m = M(v_{r+1})+\cdots+M(v_C)$. The right-hand side is a sum of odd numbers, and can be even only if the number $(C-r)$ of terms is even. (Each edge counts twice because S is *closed*.)

For each corner v, we are given that $l \leqslant M(v) \leqslant M$. Also $M \leqslant C-1$ since at most one edge can join v to any other corner (of which there are $C-1$). Thus the sum $2E$ of the numbers $M(v)$ is a sum of C terms each lying between l and $C-1$; so $lC \leqslant 3E \leqslant C(C-1)$.

***10.** Write $M(v_i)=2k_i$, where k_i is a whole number >0. Cancelling 2's gives $E=k_1+k_2+\cdots=7$ so $C\leqslant 7$ (if $C=7$, then each k_i is 1). If the various possibilities for the k's are to correspond to an actual paper surface, then by No. 9 above $2E=14\leqslant C(C-1)$ so C must be either 5, 6 or 7. Since S is closed, then $P\geqslant 2$; and by No. 8 above, $\chi \leqslant 2$. Thus if $C=7$

then $2 \leqslant P = \chi + E - C \leqslant 2$, so $P = 2$. Hence S consists of two 7-sided panels, joined along their common boundary to make an egg-shape. If $C = 6$, then some k (say k_1) is 2, and the rest are 1. But then we see from Fig. C.4

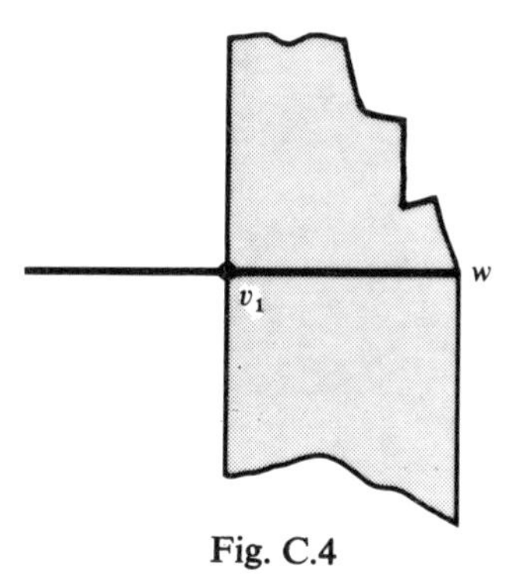

Fig. C.4

that at the corner w we cannot have $M(w) = 2$. If $C = 5$, then there are for k_1, k_2, k_3, k_4, k_5 possibilities (3, 1, 1, 1, 1) and (2, 2, 1, 1, 1). The first possibility does not correspond to a paper surface because we would have a corner like w in Fig. C.4. Nor does the second possibility, because if $M(v_1) = 4$ as in the figure, the 4 remaining vertices are as drawn there, again with the 'wrong' multiplicities for an actual surface; the remaining 3 edges cannot make them 'right'.

***13.** $C = 2E/p$ by No. 9 and $P = 2E/q$ by No. 12. Then

$$2 = \chi_S = C - E + P = \frac{2E}{p} - E + \frac{2E}{q}.$$

Cancelling the 2, we get
$$\frac{1}{p} + \frac{1}{q} = \frac{1}{2} + \frac{1}{E} > \frac{1}{2},$$

since $E > 0$. Hence, since p, q are whole numbers, we cannot have them both > 4, so p (say) is 3 or 4. If $p = 3$, then q can be 3, 4 or 5; and if $p = 4$, then q can only be 3.

For a general χ, we obtain

$$\frac{1}{p} + \frac{1}{q} = \frac{1}{2} + \frac{\chi}{E}, \quad \text{and if } \chi = 0 \text{ then } \quad \frac{1}{p} + \frac{1}{q} = \frac{1}{2};$$

by inspection (or elementary number theory) the only possibilities are $(p,q) = (3,6)$, $(6,3)$, $(4,4)$, corresponding to panellings by hexagons, triangles and rectangles.

Exercise 2.9 (p. 23)

χ is zero and one, respectively.

Exercise 3.2A (p. 25)

Use the splitting $P_1 + \cdots + P_h$ of P, mentioned earlier.

Exercise 3.3 (p. 27)

1. $\chi = -7$; for the closed surface $\chi = -6$.

2. Express a punctured sphere of genus g as $S = T + U$, where T and U are punctured spheres of genus $g - 1$ and 1, and such that they have just one edge in common. Then by Exercise 2.6, No. 7,

$$\chi(S) = \chi(T) + \chi(U) - 1 = [1 - 2(g - 1)] + (1 - 2) - 1 = 1 - 2g.$$

Exercise 3.4 (p. 28)

1. $\chi(\text{torus}) = 0$. $\chi(\text{double torus}) = -2$.

3. For the effect of adding a lid, see solution to Exercise 2.6, No. 8. If S is closed then $\beta_S = 0 = 1 - w + a - v$ so $a = w + v - 1$. Therefore

$$w - u = w - (a + b) = 1 - v - b,$$

whence

$$\chi_S = 1 + (w - u) - v = 2 - 2v - b.$$

If $\chi_S = 2$, then $2v + b = 0$; hence, since $v \geqslant 0$ and $b \geqslant 0$ then $v = b = 0$, i.e. no ear is twisted and there are no bridges. Also,

$$u = a + b = a = w + v - 1 = w - 1.$$

When $\chi_S = 1$, then $2v + b = 1$, so $v = 0$ and $b = 1$. Hence

$$u = a + 1 = (w - 1) + 1 = w.$$

Exercise 3.5 (p. 29)

These questions are for experiment and discussion. In No. 6, however, one observes that if a surface S is orientable, and a twisted ear is added, then one of the arrows, at its ends on S, is opposed to whatever arrow we put on the ear. This argument does not work without further justification for a twisted bridge, since its ends are on different boundary curves and it is unclear to say that their arrows should go in the 'same sense' unless we use both the Internal and External Properties (p. 29). But see Remark 2, p. 31. In either case, however, $S + P$ is now non-orientable.

Exercise 3.6 (p. 31)

1. Yes, if the panels are sufficiently small. The frame's surface is then a closed surface of genus 2, but if we include the (thick) lenses, the genus changes to zero.

2. Assuming the handles are solid, the surface of the pan has genus zero, while that of the kettle (whose lid has one small hole to let out the steam) is two, because the hole and spout form one tunnel, and the handle adds another. Ordinary scissors have a surface of genus 2.

3. 0, 1, 0; the genus of the chair depends on its design. A spoon and fork have genus zero, and for an egg-whisk we add 1 for each loop in the beaters, or hole in a cog (other than where axles pass through). The salt-cellar has genus 0 (if the hole were enlarged, the shape would be like that of a bowl). Each extra hole (to make a pepper-pot) adds 1 to the genus.

Exercise 4.1 (p. 35)

1. All but one of the surfaces of the table unite to form a lamina. The remaining surface (say the table-top) forms a lid.
Nos. **2–4.** Recall Exercise 2.6, Nos. 7, 8.

Exercise 4.2 (p. 37)

Express each surface in the form (2) on p. 36.

Exercise 4.3A (p. 41)

3. If the lid M is added to the hole below E_4 (say) in Fig. 4.9, then $L \dotplus M \dotplus E_4$ is a lamina to which the remaining ears are added. If instead, M is added along the outer boundary, then $M \dotplus Q$ is a lamina (where Q is the panel shown in Fig. C.5) and then $(M \dotplus Q) \dotplus R$ is also a lamina T.

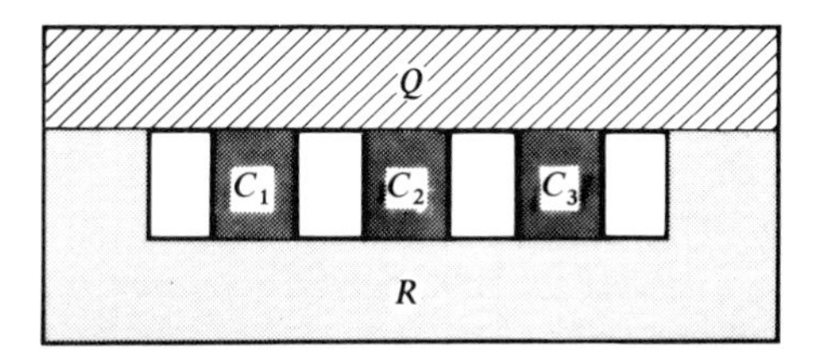

Fig. C.5

Thus $M + \mathscr{P}_4$ is $T +_e C_1 +_e C_2 +_e C_3$, where the C's are the panels shown, so $M + \mathscr{P}_4$ has $\mathscr{P}_3$ as plan. A similar argument holds for $\mathscr{P}_n$, which would have $n-1$ ears corresponding to the C's.

4. $\beta(\mathscr{P}_n) = 1 + n$; $\chi(\mathscr{P}_n) = 1 - n$ (here $C = 6n + 4$, $E = 8n + 4$, $P = n + 1$).

Exercise 4.3B (p. 42)

1. $\beta(\mathscr{T}_4) = 1$, $\chi(\mathscr{T}_4) = -7$, $\beta(\mathscr{T}_n) = 1$, $\chi(\mathscr{T}_n) = 1 - 2n$.

Exercise 4.3C (p. 43)

2. The rim and hub each have a torus as their surface. The visible surface of a spoke is an annulus, so the surface of rim + hub + one spoke is a sphere with 2 handles. The remaining 15 spokes each add one more handle.

3. 0, 0, 1 (because of the digestive tract), 0; regarding a skull as a thick 'helmet' with holes for eyes, ears, nose and mouth, the surface is a sphere with 6 handles. There are more holes if we take into account the extremely complex physiological details.

4. Using Exercise 3.3, No. 2 and Exercise 2.6, No. 8, $\chi = 1 - 2g + 1 = 2 - 2g$.

Exercise 4.4A (p. 44)

1. $\mathscr{S}_{3,2,0}$, $\mathscr{S}_{3,0,1}$. The third surface is an annulus with 4 extra holes, two twisted ears and one handle, with plan $\mathscr{S}_{5,2,1}$.

2. $\beta(\mathscr{S}_{1,2,3}) = 2$, $\beta(\mathscr{S}_{3,2,1}) = 4$,
$\chi(\mathscr{S}_{1,2,3}) = -8$, $\chi(\mathscr{S}_{3,2,1}) = -6$.

3. Use Exercise 2.6, Nos. 3, 4, and the fact that if a twisted ear is added, there is no change in β. Adding a handle does not change β since an untwisted ear increases β by 1 and the bridge of the handle cancels out this increase.

6. Draw one arrow clockwise in each panel of Fig. 4.14 (with no twisted ear present).

Exercise 4.4B (*p. 45*)

1. We can split $\mathscr{S}_{p,q,r}$ into a sum $\mathscr{S}_{p,q,0} \dotplus \mathscr{S}_{0,0,r}$ by dividing the plan illustrated in Fig. 4.14 into two, using a line that separates the handles from the rest. On the other hand, if we join $\mathscr{S}_{p,q,0}$ to $\mathscr{S}_{0,0,r}$ along *any* arc to form a sum of the form $\mathscr{S}_{p,q,0} \dotplus \mathscr{S}_{0,0,r}$, we must show that the result is in Family ($\mathscr{S}_{p,q,r}$). By introducing a narrow panel along the arc, like the black regions in Fig. 4.9(*b*), we can ensure that the arc is in the boundary of M and of N, where $\mathscr{S}_{p,q,0} = M + \text{ears}$, $\mathscr{S}_{0,0,r} = N + \text{handles}$, and M, N are laminas. Thus $\mathscr{S}_{p,q,0} \dotplus \mathscr{S}_{0,0,r} = (M \dotplus N) + \text{ears} + \text{handles}$, $\sim \mathscr{S}_{p,q,r}$, since $M \dotplus N$ is a lamina.

The second part is done similarly, and of course it is easy to separate the twisted ears on $\mathscr{S}_{p,q,r}$ from the rest, to get a splitting of $\mathscr{S}_{p,q,r}$ into the sum $\mathscr{S}_{p,0,r} \dotplus \mathscr{S}_{0,q,0}$. Just as before, however, one must show that *any* such sum lies in Family ($\mathscr{S}_{p,q,r}$).

2. The number β counts the number of boundary curves, which is independent of the way in which we group the panels of the surface S to form the laminas of the prescription (8). Now consider χ, and let L be one of the laminas; L is a sum of panels $P_1, P_2, \ldots, P_k$, say, and as we have seen before, $\chi(L) = 1$. Let the numbers of corners, edges and panels of S be C, E, P, and let these numbers for L (panelled by the P's) be c, e, k. If the boundary of L has x corners, it has x edges, so the numbers of corners and edges *not* inside L are $C-(c-x)$, $E-(e-x)$; and if we think of S as panelled by L and those panels not in L, we compute an Euler number

$$[C-(c-x)]-[E-(e-x)]+(P-k+1) = \chi(S)-(c-e+k)+1 = \chi(S),$$

since $c-e+k = \chi(L) = 1$. Thus, we get the same number $\chi(S)$ if we erase all edges and corners inside L. We now do a similar erasure on each lamina in turn, without altering the number $\chi(S)$. Finally, we are left with S divided up into the laminas, but each now counting as a single panel. Again we compute the Euler number, and again it is $\chi(S)$ as before.

Exercise 4.5 (*p. 48*)

For brevity, write $S \sim T$ to mean that Family (S) = Family (T). Then

$$\begin{aligned} \mathscr{S}_{0,4,0} &= \mathscr{S}_{0,3,0} \dotplus M \quad (M \text{ a Moebius band}),\\ &\sim \mathscr{S}_{0,1,1} \dotplus M \quad (\text{by the Trading Theorem}),\\ &\sim \mathscr{S}_{0,2,1}. \end{aligned}$$

Similarly

$$\mathscr{S}_{0,5,0} = \mathscr{S}_{0,4,0} \dotplus M \sim \mathscr{S}_{0,2,1} \dotplus M, \quad \text{using the previous result}$$
$$\sim \mathscr{S}_{0,3,1}.$$

But

$$\mathscr{S}_{0,3,1} \sim \mathscr{S}_{0,3,0} \dotplus T \quad (T \text{ a punctured torus}),$$
$$\sim \mathscr{S}_{0,1,1} \dotplus T \quad (\text{by the Trading Theorem}),$$
$$\sim \mathscr{S}_{0,1,2}.$$

In general,

$$\mathscr{S}_{0,2n,0} \sim \mathscr{S}_{0,2,n-1} \quad (\text{if } n \geqslant 1)$$

and

$$\mathscr{S}_{0,2n+1,0} \sim \mathscr{S}_{0,1,n}.$$

The proof is by induction on n, because both statements hold if $n = 1$, the second by the Trading Theorem. The principal inductive step is:

$$\mathscr{S}_{0,2n+1,0} \sim \mathscr{S}_{0,2n,0} \dotplus M \sim \mathscr{S}_{0,2,n-1} \dotplus M \quad (\text{by inductive hypothesis}),$$
$$\sim \mathscr{S}_{0,3,n-1} \sim \mathscr{S}_{0,3,0} \dotplus \mathscr{T}_{n-1},$$
$$\sim \mathscr{S}_{0,1,1} \dotplus \mathscr{T}_{n-1} \quad (\text{Trading Theorem}),$$
$$\sim \mathscr{S}_{0,1,n},$$

while

$$\mathscr{S}_{0,2n,0} \sim \mathscr{S}_{0,2n-1,0} \dotplus M \sim \mathscr{S}_{0,1,n-1} \dotplus M \quad (\text{inductive hypothesis}),$$
$$\sim \mathscr{S}_{0,2,n-1}.$$

Exercise 4.6 (p. 54)

1. $\mathscr{S}_{q,0,r} \sim \mathscr{P}_q \dotplus \mathscr{S}_{0,0,r}$; but $\mathscr{P}_q$ is a sum of annuli as indicated by $\mathscr{P}_3 = A_1 \dotplus A_2 \dotplus A_3$ in Fig. C.6. If the hole in each A_j is filled by a Moebius

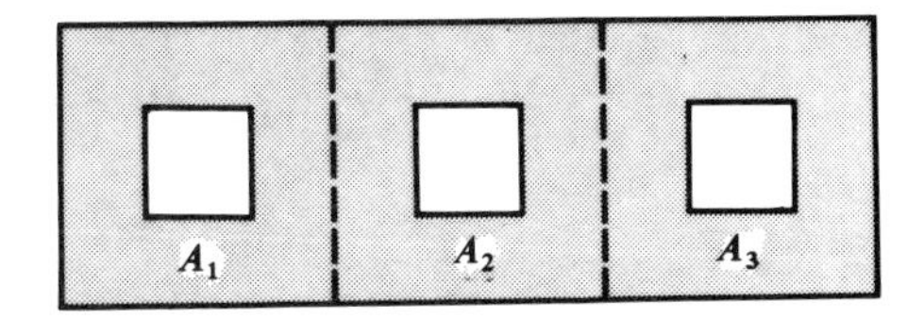

Fig. C.6

band M_j, then $\mathscr{P}_3$ plus the 3 Moebius bands is

$$(A_1 + M_1) \dotplus (A_2 + M_2) \dotplus (A_3 + M_3).$$

But $A_j + M_j$ is simply a broader Moebius band than M_j, so

$$\mathscr{P}_3 + M_1 + M_2 + M_3 \sim \mathscr{S}_{0,3,0}.$$

Similarly, if q bands M_j are added to the q holes of $\mathscr{P}_q$, the sum is $\mathscr{S}_{0,q,0}$. Hence

$$\mathscr{S}_{q,0,r} + M_1 + \cdots + M_q \sim (\mathscr{P}_q + M_1 + \cdots + M_q) \dotplus \mathscr{S}_{0,0,r},$$
$$\sim \mathscr{S}_{0,q,0} \dotplus \mathscr{S}_{0,0,r} \sim \mathscr{S}_{0,q,r}.$$

3. See p. 27 and Section 2.8.

Exercise 5.1 (p. 58)

1. It follows from the solution to Exercise 4.5, that (using the symbol ~ as there)

$$\mathscr{S}_{p,q,r} \sim \mathscr{S}_{p,0,r} \dotplus \mathscr{S}_{0,q,0} \sim \mathscr{S}_{p,0,r} \dotplus \mathscr{S}_{0,1,t},$$

where $t = \frac{1}{2}(q-1)$ since q is odd

$$\sim \mathscr{S}_{p,1,r+t} = \mathscr{S}_{p,1,u}.$$

If q had been even, say $q = 2t$, then $\mathscr{S}_{0,q,0} \sim \mathscr{S}_{0,2,t-1}$, so

$$\mathscr{S}_{p,q,r} \sim \mathscr{S}_{p,0,r} \dotplus \mathscr{S}_{0,2,t-1} \sim \mathscr{S}_{p,2,r+t-1}, \text{ and } v = r - 1 + \tfrac{1}{2}q.$$

2. $\mathscr{S}_{p,q+2r,0} \sim T \dotplus \mathscr{S}_{0,3,0}$, $T = \mathscr{S}_{p,q-1+2r-2,0}$ $(q \geqslant 1)$
$\sim T \dotplus \mathscr{S}_{0,1,1}$ (by the Trading Theorem)
$\sim \mathscr{S}_{p,q+2r-2,1}$
$\sim \mathscr{S}_{p,q+2r-4,2}$ (similarly)
$\sim \cdots \sim \mathscr{S}_{p,q,r}$.

Exercise 5.3 (p. 60)

1. $b = 1 + p = 2$, so $p = 1$; $c = -21 = 1 - (p + q + 2r)$, so $21 = q + 2r$. Hence any plan will do, of the form $\mathscr{S}_{p,q,r}$ provided $q + 2r = 21$; so the possibilities for (q,r) are $(1,10)$, $(3,9)$, . . ., $(21,0)$.

2. This follows from the Conclusion (p. 56) and the Addition Equations.

3. Let the octagonal panels be L, M, and let P_1, P_2, P_3, P_4 be the twisted panels. Then $A = (L \dotplus P_1) + (M \dotplus P_2)$ is a 2-twisted annulus, with plan $\mathscr{S}_{1,0,0}$ (recall our intention on p. 10 to ignore an even number of twists in an annulus). When P_3 is added to A, it joins together the two boundary curves, as an untwisted bridge, by Remark 2, p. 31. P_4 is therefore an ear, untwisted according to the Test on p. 17. By the Addition Equations, then,

$$A + P_3 \sim \mathscr{S}_{1,0,0} +_{\mathrm{b}} P_3 \sim \mathscr{S}_{0,0,1},$$

so

$$(A \dotplus P_3) + P_4 \sim \mathscr{S}_{0,0,1} +_{\mathrm{e}} P_4 = \mathscr{S}_{1,0,1}.$$

Thus the surface is a torus with two holes.

Exercise 6.1 (p. 64)

2. Let the numbers of corners, edges and panels in the dual be c, e, p. Since there is exactly one dual corner in each original panel, we have $c = P$. Each dual edge $v_i v_j$ can be drawn to cut the common edge of panels P_i, P_j exactly once, so $e = E$. Each corner X of the original panels is the centre (see Fig. C.7) of a lamina L formed by all the panels possessing X as corner, so X lies inside the Jordan curve K formed by the dual edges $v_i v_j$, where P_i, P_j are in L. K surrounds a dual panel Q in L, containing X in its interior. Hence $p = C$.

3. If X is chosen to represent Q for each corner and dual panel (as in Fig. C.7) then the dual of the dual panelling can be taken to be the original.

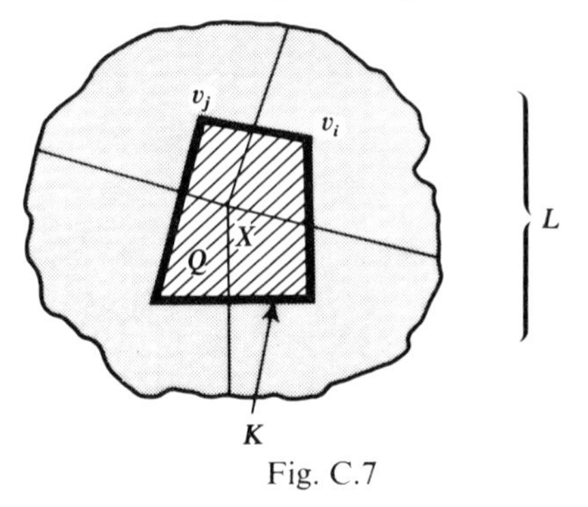

Fig. C.7

Exercise 6.6 (*p. 73*)

5. If we fill in the hole, we obtain a planar map, so 5 colours suffice. If 4 would not suffice for the annulus, then they would not suffice when the hole is filled in, so we would have a counter-example to the Four-Colour Conjecture.

Exercise 7.1 (*p. 76*)

4. D must be a lamina, otherwise there would be a boundary curve with a crossing point like a figure eight (or worse).

Exercise 8.2 (*p. 90*)

2. If the polyhedron were to rest on an edge or face, we would have a curve or plateau of minima, contrary to the rule for being properly arranged.

3. $\chi = -2$ and there are at least one pit and one peak among the lowest and highest points of the surface. Therefore by the mountaineer's equation, $-2 \geqslant 1 -$ passes $+ 1$, so there are at least 4 passes. With the pits and peaks, this makes at least 6 critical points.

Exercise 8.3 (*p. 94*)

1. His route is a winding Jordan curve that surrounds a tract T of country. Imagine that the whole tract T is raised up from sea-level, on a pillar. The vertical surface plus T is then a lamina, so $\chi = 1 = b - c + d$.

2. There is at least one pit and one peak so the number of passes is (pits + peaks $- \chi) \geqslant (1 + 1 - 0) = 2$.

Miscellaneous exercises (*p. 95*)

1. If the numbers of corners, edges and panels for A, B, C are $c, c', c'', p, \ldots$, etc., then $p'' = p + p'$, $e'' = e + e' - n$, $c'' = c + c' - n$, where $n =$ no. of edges on the boundary of A that is taped to B. Hence $\chi(C) = c'' - e'' + p''$ $= \chi(A) + \chi(B)$. Clearly, the boundary of C consists of those boundary curves of A and B that are not involved in the taping, hence the equation for $\beta(C)$.

2. $\beta(A) = p + 1$, $\beta(B) = u + 1$, so, by No. 1, $\beta(C) = u + p > 0$ since C is not closed. Hence C has a plan of the form $\mathscr{S}_{x,y,z}$, with $x = \beta(C) - 1 \geqslant 0$. Also $\chi(A) = 1 - (p + q + 2r)$, by Exercise 4.4A, No. 3, and similarly for B; so, by No. 1,

$$\chi(C) = 2 - [p + u + q + v + 2(r + s)].$$

To determine y and z, see p. 60: we first form $\beta(C) + \chi(C) = \gamma$ which here has the parity of $q + v$. Suppose first that $q = v = 0$. Then we can orient each of A and B and thus orient C, so $y = 0$. Hence $z = r + s$ (see p. 60). If q and v are not both zero, we may assume that each lies between 0 and 2, and also that $y = 1$ or $y = 2$ according as γ is odd or even. If $y = 1$, then (see p. 60) $z = \frac{1}{2}(1 - \gamma)$ and if $y = 2$ then $z = -\frac{1}{2}\gamma$.

3. If $W = V - T$, then W is a punctured version of S so $W \sim \mathscr{S}_{0,q,r}$. We can join T to W along an arc and then add a lid L to close up the hole thus formed. Hence

$$V \sim (W \dotplus T) \mathring{+} L \sim (\mathscr{S}_{0,q,r} \dotplus \mathscr{S}_{0,0,1}) \mathring{+} L = \mathscr{S}_{0,q,r+1} \mathring{+} L = \mathscr{C}_{q,r+1}.$$

4. $\mathscr{C}_{1,r}$ is obtained by adding one Moebius band M and r handles to a sphere S with one hole (see p. 54). But $S \sim$ panel, so $M + S = M +$ lid $=$ real projective plane R. Thus $\mathscr{C}_{1,r}$ can be obtained by adding r handles to R. Similarly, for $\mathscr{C}_{2,r}$, we can add r handles to $B = (A + M_1) + M_2$, where A is a sphere with two holes (i.e. an annulus) and the M's are Moebius bands. But $A = U +_e V$, where U, V are laminas, so we may write $B = [(M_1 \dotplus U) \dotplus M_2] \mathring{+} V = K +$ lid, $K =$ punctured Klein bottle (see Fig. 2.7); thus B is a Klein bottle.

5. Let L denote a lid. The surface is $(\mathscr{P}_3 + L) + 3$ Moebius bands, $\sim \mathscr{S}_{0,3,0} + L$ by Exercise 4.6, No. 1 (see p. 117). Now $\mathscr{S}_{0,3,0} \sim \mathscr{S}_{0,1,1}$ by the Trading Theorem, so the surface is $\mathscr{S}_{0,1,1} + L = \mathscr{C}_{1,1}$.

6. As in No. 5, above, the surface is $(\mathscr{P}_{q+2k} + L) + (q + 2k)$ Moebius bands $\sim \mathscr{S}_{0,q+2k,0} + L \sim \mathscr{S}_{0,q,k} + L = \mathscr{C}_{q,k}$, using the Trading Theorem, as on p. 57.

7. $\chi(S + L) = \chi(S) + 1$; $\chi(\mathscr{S}_{p,q,r}) = 1 - (p + q + 2r)$ (see Exercise 4.4A, No. 3). Hence $\chi(\mathscr{C}_{1r}) = \chi(\mathscr{S}_{0,1,r} + L) = [1 - (1 + 2r)] + 1 = 1 - 2r$. Similarly if $q = 2$. Now see No. 6.

8. (*a*) Let M denote a Moebius band. Then using No. 1 we have

$$\chi(S_1) = \chi(M) + 1 = 1, \quad \chi(T_1) = \chi(S_1) - 1 = 0;$$
$$\chi(S_2) = -1, \quad \chi(T_2) = \chi(S_2) - 1 = -2.$$

(*b*) $\chi(S) = \chi(T_1) + \chi(T_2) = -2$, and S has one boundary curve (that of S_2). Since S contains M, it is non-orientable. Thus $S \sim \mathscr{S}_{p,q,r}$, $q \neq 0$, where p, q, r can be computed as on p. 61. Here $b = 1$, $c = -2$ so $b + c$ is odd and $q = 1$ (see p. 60); thus $p = 0$, $r = 1$ (p. 61): $S \sim \mathscr{S}_{0,1,1}$.

9. Exactly similar calculation to that of No. 6. Here

$$\chi(S_1) = -1, \quad \chi(T_1) = -2, \quad \chi(S_2) = 0, \quad \chi(T_2) = -1,$$
$$\chi(S) = -2 + (-1) = -3.$$

S has one boundary curve, so $b = 1$, $c = -3$, $b + c$ is even and hence $S = \mathscr{S}_{0,q,r}$ with $q = 2$ (p. 60). Therefore $r = -\frac{1}{2}(b + c) = 1$; $S \sim \mathscr{S}_{0,2,1}$.

10. $\beta(\mathscr{P}_4) = 5$, so $\beta(S) = 4$ and $\chi(S) = \chi(\mathscr{P}_4) - 1 = -3$ (Exercise 2.6, No. 4).
(*a*) Since $\mathscr{P}_4 = \mathscr{S}_{4,0,0}$ then $S = \mathscr{P}_4 +_{\text{tb}} P \sim \mathscr{S}_{3,2,0}$ by the Trading Theorem.

11. Let T_k denote the surface formed by adding the $4 + k$ twisted panels to the 'top' and 'bottom' $(8 + 2k)$-gons. Then T_k is *orientable*; to see this when k is fairly large, using orienting arrows, is complicated, so we use another argument. Suppose all panels are black on one side, white on the other, and tape the panels to be twisted, first to the top polygon, in such a way that their white sides continue that of the top. Call the resulting lamina D. Now lay down the bottom polygon B white-side up; and hold D above it, also white-side up. Each 'spoke' of D must be twisted once and its end taped to an edge of B. The twist causes the black surface of the spoke to pass onto that of B. After L is thus joined to B we obtain T_k with *two* sides, one black, one white. Hence T_k is orientable. Next, by labelling the corners of the top and bottom consecutively as $v_1, v_2, \ldots, v_{8+2k}, w_1, w_2, \ldots, w_{8+2k}$, and considering the route $v_1 w_2 w_3 v_4 v_5 w_6 \ldots$, we find that it returns to v_1 either by visiting *all* the v's and w's, or just half of them. The first alternative occurs when k is odd (and then $\beta(T_k) = 1$) and the second when k is even (and then $\beta(T_k) = 2$). Hence, if we add a new twisted panel P to T_k in order to form T_{k+1}, then P is a bridge if k is even, and an ear if k is odd; in neither case is P a twisted bridge or twisted ear because T_{k+1} is orientable. We saw in Exercise 5.3, No. 3 that $T_0 \sim \mathscr{S}_{1,0,1}$, so $T_1 \sim T_0 +_{\text{b}} P \sim \mathscr{S}_{0,0,2}$, $T_2 \sim T_1 +_{\text{e}} P \sim \mathscr{S}_{1,0,2}$, and by induction on k,

$$T_k \sim \mathscr{S}_{h-1,0,x+2} \quad (k \geqslant 1),$$

where $k = 2x + h$ and $h = 1$ or 2. (E.g. $T_{27} \sim \mathscr{S}_{0,0,15}$, $T_{56} \sim \mathscr{S}_{1,0,29}$.)

12. S_k is T_{k-4} in the notation of No. 11. Therefore $S_5 = T_1 \sim \mathscr{S}_{0,0,2}$, $S_4 = T_0 \sim \mathscr{S}_{1,0,1}$, and when they are joined along a common boundary, the resulting surface V is orientable. Thus, by No. 1, $V \sim \mathscr{S}_{0,0,3}$. If S_4 and S_5 are replaced by S_p, S_q, we use No. 11 similarly, looking at the separate cases when p and q are both odd (so $\beta(S_p) = 1 = \beta(S_q)$ and V is closed), and when one at least is even. The resulting formulae are fairly clumsy to write down in general, although easy enough for specific values of p and q.

13. (*a*) See p. 53. (*b*) $\mathscr{S}_{p,2,r} \mathbin{\mathring{+}} M$ is given by No. 2 with $B = \mathscr{S}_{0,1,0} (\sim M)$, so $\beta(C) = p$, $\chi(C) = -(1 + p + 2r)$. Thus $b + c$ (p. 60) $= -(1 + 2r)$ which is odd.
Therefore $C \sim \mathscr{S}_{p-1,1,1+r}$ (see p. 61).

14. If S is assembled from panels as $S = P_1 + P_2 + \cdots + P_n$ then its boundary is clearly a Jordan curve if $n = 1$. Now use induction on n. The boundary of $T = P_1 + \cdots + P_{n-1}$ consists of Jordan curves, so that of S consists of some of those, together with modifications of the rest arising from the addition of P_n as ear, bridge or lid. These modifications clearly produce Jordan curves.

15. If the panel P has n corners, then it has n edges. The barycentric subdivision has $2n + 1$ corners, $4n$ edges and $2n$ panels. Therefore the Euler number of the new subdivision is $(2n + 1) - 4n + 2n = 1$ as before.

16. Let C', E', P' be the new numbers for the subdivision. Then $C' = C + E + P$ since there is one new corner for each old edge and panel. For each old panel Q_j, let q_j denote the number of its edges. Then in the subdivision, Q_j is replaced by $2q_j$ triangles, so $P' = 2\ \Sigma q_j$; and each old edge is divided into two while $2q_j$ new edges are introduced in the interior of Q_j, so $E' = 2E + 2\ \Sigma q_j$. Thus $C' - E' + P' = C - E + P$ as before. (Note: by Exercise 2.6, No. 11, $\Sigma q_j = 2E$, so $P' = 4E$, $E' = 6E$. However, since the Σ's cancel above, they need not have been summed over all panels, i.e. we would have the same value for χ even if we had failed to subdivide some of the panels.)

17. First subdivide S barycentrically (No. 15) to ensure that no edge has both ends on J unless it lies in J, and no panel has more than one edge on J. Number the corners of J consecutively as $v_1, v_2, \ldots, v_n$; then the edge $v_n v_1$ is free and lies on just one panel P_1. Let e_1 be the other edge of P_1 with corner v_1; by the subdividing process, e_1 is not on J, and so lies on a second panel P_2, whose next edge after e_1 is e_2. If P_2 has an edge on J, then e_2 is $v_1 v_2$, but if not, then we find a panel P_3 with corner v_1 and edges e_2, e_3 in the same way. Continuing in this way round J, we label consecutively all the edges having just one end on J, as $e_1, e_2, \ldots, e_m$ $(m \geqslant n)$. Next introduce a new vertex w_i at the mid-point of e_i, $i = 1, \ldots, m$, and join w_i to w_{i+1} by a new edge to split off from the panels P_j new, smaller, panels Q_j that have an edge or vertex in J. The required neighbourhood is then the sum of these panels Q_j, and its two boundary curves consist of J and $w_1 w_2, \ldots, w_n w_1$.

18. A suitable sketch could have a feature like the two horns on a cow's head: between the horns lies a saddle point.

19. In the notation of p. 93, $m - s + M = \chi(S) = -2$ so $s = m + M + 2 \geqslant 4$ since m and M are always at least 1. Hence the number of critical points is $s + m + M \geqslant 4 + 2 = 6$.

20. An exposition is lengthy: see Griffiths, H. B. 'The Topography of 2×2 real matrices, III', *International Journal of Mathematical Education in Science and Technology*, **10** (1979), 603. But the horizontal sections are hyperbolae with vertices on a 'figure eight' curve, its loops in different planes. Once this is verified, a sketch can be drawn.

21. Here, we have the same edge-labels as before, but c is now taped to u'' and v' to a. Thus the new table differs from that on p. 97, in that the v'-column is now headed by c, and the top 1 and lowest 0 are now interchanged.

22–25. It probably suffices to deal with No. 24. We first pick out the free edges e_2, e_4, e_7, e_8 – distinguished by having only a single 1 under the row of panels in the table. The corners of e_2 are v_2, v_3 (indicated by the 1's under the v's), and the corners of the other free edges show that there is one boundary curve $e_2 e_7 e_4 e_8$ with corners $v_2 v_3 v_4 v_1$ (by inspection). Hence the surface S has $\beta(S) = 1$, while $\chi(S) = 4 - 8 + 3 = -1$ so $b + c = 0$ (p. 60); therefore $q = 0$ or 2, and then either $S \sim \mathscr{S}_{0,0,1}$ or $S \sim \mathscr{S}_{0,2,0}$. We must therefore decide directly whether or not S is orientable. The table shows that P_1 has edges e_1, e_3, e_5, e_6 corresponding to the 1's in its column, so the boundary of P_1 consists of e_1 $(= v_2 v_4)$, e_7 $(= v_4 v_3)$, e_3 $(= v_3 v_1)$, e_8 $(= v_1 v_2)$, in that order. Since e_5, e_6 are edges of P_2 and its other edges are e_2, e_4, then P_2 is added to P_1 as an ear. Now e_5 has ends v_2, v_3, while e_6 has ends v_1, v_4 so a sketch shows that P_2 must be twisted if we are to join it to P_1. Thus S contains $P_1 +_{\text{te}} P_2$ which is a Moebius band; hence S is non-orientable so q above is 2. Therefore $S \sim \mathscr{S}_{0,2,0}$.

The tables in Nos. 23 and 25 are treated similarly. Neither represents a surface with free edges; the first represents a tetrahedron (in the usual panelling), and the second a torus.

Note: if a surface S can be oriented, then it cannot contain a Moebius band M, otherwise the orienting arrows of S would also orient M. Thus, *if a surface contains M, it cannot be orientable* – a condition we used above without comment.

References

(The more advanced texts are starred.)

*1. Ahlfors, L. V. et al. Papers in *Analytic Functions*, Princeton University Press, Princeton (1960).

*2. Ahlfors, L. V. and Sario, L. *Riemann Surfaces*, Princeton University Press, Princeton (1960).

*3. Arnold, V. I. and Avez, A. *Ergodic Problems of Classical Mechanics*, Benjamin, New York (1968).

4. Armitage, J. V. and Griffiths, H. B. *A Companion to Advanced Mathematics, I*, Cambridge University Press, London (1969).

5. Courant, R. and Robbins, H. *What is Mathematics?*, Oxford University Press, London (1941).

6. Coxeter, H. S. M. *Introduction to Geometry* (Second Edition), Wiley (1969).

7. Griffiths, H. B. and Hilton, P. J. *A Comprehensive Textbook of Classical Mathematics: a contemporary interpretation*, Springer (1978).

8. Griffiths, H. B. and Howson, A. G. *Mathematics: Society and Curricula*, Cambridge University Press, London (1974).

9. Hilbert, D. and Cohn-Vossen, S. *Geometry and the Imagination*, Chelsea, New York (1952).

10. Hilton, P. J. (Editor). *Studies in Modern Topology* (Mathematical Association of America), Studies in Mathematics, vol. 5. Buffalo, New York (1968).

11. Hirst, K. E. and Rhodes, F. *Conceptual Models in Mathematics* (Mathematical Studies No. 5), George Allen and Unwin, London (1971).

12. Klein, F. *On Riemann's Theory of Algebraic Functions and Their Integrals.* Dover Reprint, New York (1963).

*13. Milnor, J. *Morse Theory*, Annals of Mathematics Studies, 51, Princeton University Press, Princeton (1963).

14. Newman, M. H. A. *Elements of the Topology of Plane Sets of Points*, Cambridge University Press, London (1954).

15. Ore, O. *Graphs and Their Uses*, Random House, New York (1963).

16. Saaty, T. L. 'Thirteen colorful variations on Guthrie's four-color conjecture', *American Mathematical Monthly*, **79** (1972), pp. 2–43.

17. Seifert, H. and Threlfall, W. *Lehrbuch der Topologie*, Chelsea, New York (1947).

18. Wang, H. 'Towards Mechanical Mathematics', *IBM Journal of Research and Development*, **4** (1960).

*19. Wilder, R. L. *Topology of Manifolds*, American Mathematical Society Colloquium Publications, 32 (1949).

20. Willmore, T. J. *An Introduction to Differential Geometry*, Clarendon Press, Oxford (1959).

21. Wilson, R. J. 'Map-colouring problems', *Bulletin of the Institute of Mathematics and Its Applications*, **16** (1980), pp. 16–19.

Index of symbols

(Page numbers here refer to first appearance, and teaching note if any)

$\mathscr{C}_{q,r}$ family of closed surfaces, 54
$S \sim T$ *S* in the same family as *T*, 41, 96
$\mathscr{P}_n$ standard planar region with *n* holes, 39, 101
$\mathscr{S}_{p,q,r}$ standard paper surface, 44, 101
$\mathscr{T}_n$ standard punctured torus with *n* handles, 42, 101
Ⓣ reference to Appendix B, x
β number of boundary curves, 18, 100
χ Euler number, 18,100

$+$ addition of surfaces, 399
$\dot{+}$ panel added along arc, 25
$\overset{\circ}{+}$ panel added as a lid, 25
$+_e$ panel added as an ear, 35
$+_{te}$ panel added as a twisted ear, 37
$+_b$ panel added as a bridge, 36
$+_{tb}$ panel added as a twisted bridge, 38
$+_x$ panel added in any one of previous ways, 54

Subject index

absolute (maximum, minimum), 89
addition equations, 57, 59
Addition Theorem, 48, 57
Agreement 1, 32
Agreement 2, 33, 100
Agreement 3, 34, 100
Agreement 4, 39, 100
Agreement 5, 39, 45
Agreement 6, 53, 101
Agreement 7, 67, 100
A-level, 106, 108
amoeba, 43, 115
analysis, case-by-case, 106
analysis (subject), 101
annulus, 9, 35, 69, 73, 76, 103
arc, 22, 67, 101
arc succession, 25, 32
arrow (orienting), 28, 44, 79, 114
assembly, ordered, 75
Assembly Theorem, 76, 94, 104
associativity, 106
axioms, ix, 104

bar, 83, 86
bay, 39
beetle, 17
Big Question, 30, 58
bijection, 105, 106
boundary, 7, 81, 104
bridge, 17, 50; multi-, 25; twisted, 17, 52

calculus, 105, 108
Cayley, 82
characteristic, 18, 104
Classification Theorem, 61
closed surface, 7, 54, 61, 94
cocoon, 34, 66, 69, 73, 105
colouring problems, 69, 73
compact, 100, 108, 110
complex (paper), 64, 76, 86
connected, 76, 100
contour, 85, 86
counter-example, 107
corner, 7
critical level, 89, 91
critical point, 85, 89
cross-cap, 54, 61
crystal, 6
cube, 7, 64
curve (Jordan), 10, 14, 86, 99, 104, 107
cylinder, 9

definitions, 34, 103
differential invariant, 66
dodecahedron, 8, 64
double torus, 27, 111
dual, 64
duality (Alexander), 104

ear, 15, 48; multi-, 24; twisted, 16, 48
earthworm, 43, 115
edge, 7; free, 7
equation: mountaineer's, 82; of hyperboloid, 110; of surface, 12; of torus, 12, 110; quadratic, 64, 107
Euler, 18, 66
Euler formula, 66
Euler number, 18, 59, 65, 72, 82, 107, 114
examinations, 107, 108
experiment, 46, 106

Family, 32, 58, 100
finer, 71
Five-colour Theorem, 69
flange, 36
Flood, 83
flow chart, 81, 108
fount (of type), 105
Four-colour Conjecture, 69
free (edge), 7
Fundamental Theorem, 58

Gallic Plan, 61
Gallic Theorem, 58
genus, 27, 103
graph theory, 7, 103

handle, 27, 42, 51, 61
Heawood's Theorem, 73
height, 82, 94, 108
homeomorphism, 99, 100, 109, 110
hyperboloid, 110

icosahedron, 8, 64
induction, mathematical, 59, 68, 105, 106
invariance, 63
Invariance Theorem, 65, 94
invariant, 65, 75, 81

isomorphism, 105
isthmus, 83, 86

Jordan, 10
Jordan curve, 10, 14, 86, 99, 104, 107
Jordan Curve Theorem, 99, 104

Klein, 21
Klein bottle, 54, 73

lamina, 32, 68
left-handed, 10, 41, 103
level, 85, 89
lid, 21, 26
linear graph, 7, 103

map, 69
manifold, 102; generalized, 100
mathematical induction, 59, 68, 105, 106
mathematical language, 101, 105
mathematical model, modelling, 99, 100, 108
mathematical surface, 99
maximum, 85, 89
Maxwell, J. C., 82
meaning, ix
minimum, 85, 89
Moebius, 10
Moebius band (or strip), 10, 37, 44, 54, 69, 73, 99, equation, 13
model, 106
Morse, M., 82
Morse, S. F. B., 82
Morse Theory, 82, 102
mountaineer's equation, 82
multiplicity, 25
mystery surface, 76, 100

non-orientable, 29, 61, 79, 94
number: Euler, 18, 59, 65, 72, 82, 107, 114; orientability, 60, 66, 75, 80; real, 22, 99, 107

octahedron, 8, 64
ordered assembly, 75
order of assembly, 62, 63, 75
orientable, 29, 44, 79, 80, 114
orientability number, 60, 66, 75, 80
Orientability Theorem, 78
oriented, 29
ovaloid, 3
overpass, 15, 41, 103

panel, 2, 32, 66, 99
paper complex, 64, 76, 86
paper surface, 7, 94; Gallic, 61; mystery, 76, 100; non-orientable, 29, 61, 79, 94; orientable, oriented, 29, 44, 79, 80, 114; with boundary, 10

parity, 60, 107
pass, 82, 84, 89; over-, 103; under-, 103
peak, 82, 84, 89
Pedagogical Axiom, ix, 104
pit, 82, 84, 89
plan, 39, 44, 58, 105, 106
planar region (or surface), 14, 39, 41
planting a concept, 105, 108
plateau, 82, 85
Plato, 7
Platonic polyhedra, 8, 34, 64
Platonic solids, 7
polygonal Jordan Curve Theorem, 99
polyhedron, regular, 8, 76
prescription, 45
prism, 9
projective geometry, 22
projective plane, 22, 54, 73
proof (notion of), x, 31
proper panelling, 64, 78
punctured Klein bottle, 21, 38, 44
punctured projective plane, 22
punctured sphere with *g* handles, 26, 27
punctured torus, 21, 37

quadratic equation, 64, 107
Question (Big), 30, 58

Recognition Claim, 54
refinement, 71
regular polyhedron, 8, 76
repanelling, 25, 63, 65, 71
ridge, 85
Rule 1, 2
Rule 2, 6

saddle point, 85, 86
Schoenflies Theorem, 99, 104, 107
screwdriver rule, 30
sculpture, xi, 56, 85
sketching, 106
skull, 43, 115
slicing plane, 90,
sphere, 109
sphere, punctured, 26, 27
sphere, with g handles, 27, 43, 54
spiral approach, 105
Star, 76, 100
subtraction (of panel), 64
surface: mathematical, 99; mystery, 76, 100; paper, 7; planar, 14

tape, 2, 65, 100
Tests, 17, 64
tetrahedron, 7, 64
theorem (explanation), 34
Theorems: Addition, 48, 57; Assembly, 76, 94, 104; Classification, 61; Five-colour, 69; Fundamental, 58; Gallic, 58; Heawood's, 73; Invariance, 65, 94;

Theorems—*cont:*
  Jordan Curve, 99, 104; Orientability, 78; Schoenflies, 99, 107; Trading, 46, 54
torus, 7, 73, 94, 110; double, 27, 111; equation, 12, 110; punctured, 21, 37
Trading Theorem, 46, 54
triangulation, 67, 100
trick, 51, 106
twist, 10, 17, 41, 80, 81, 103
two-sided, 29, 30

underpass, 15, 41, 103
understanding, 107
uniqueness (of plan), 60

vertices, 99, 104
volcano, 82, 85